HOLT
Science

Forensics and Applied Science Experiments Teacher Edition

HOLT, RINEHART AND WINSTON
A Harcourt Education Company

Austin · New York · Orlando · Atlanta · San Francisco · Boston · Dallas · Toronto · London

Credits

Lab Writer

James R. Hurley
Director of Development, American Academy of Forensic Sciences
Shell Rock, Iowa

Introduction Writers

Gary Harris
Science Writer
Albuquerque, New Mexico

Robert Davisson
Science Writer
Albuquerque, New Mexico

Safety Reviewer

Robert S. Davis
Former Science Curriculum Assistance Specialist,
State of Alabama Department of Education
Mobile, Alabama

ISBN 0-03-036793-X

4 5 6 170 09 08 07 06

Table of Contents

Program Overview. 1a
Ordering Materials 1b
Calculator-Based Laboratory Information 1c

Laboratory Safety. 1
Laboratory Techniques 8

Forensic Science
An Introduction to Forensic Science 11

Density Experiments
A Lesson on Density. 17
The Hit and Run (Archimedes' principle) **(Inquiry)**. 23
The Parking Lot Collision (The method of suspension) **(Inquiry)** . . 27
The Sports Shop Theft (Gradient column) **(Inquiry)** 31

Chromatography Experiments
A Lesson on Chromatography. 35
The Counterfeit Drugs (Strip-paper chromatography) 39
The Athletic Rivals (Paper chromatography) 43
The Questionable Autograph (Thin-layer chromatography) 47

Spectroscopy Experiments
A Lesson on Spectroscopy 51
The Fast-Food Arson (Absorbance peaks) 59
The Untimely Death (Beer's law interpolation) 63
The Assault at the Flower Shop (Soil-settling rate curve) **(Inquiry)**. 67

Identification Experiments
A Lesson on Identification 71
The Murder and the Blood Sample (DNA analysis) 77
Blood Typing (Pre-laboratory exercise). 83
The Neighborhood Burglaries (Blood typing) **(Inquiry)** 89

Environmental Chemistry

An Introduction to Environmental Chemistry 93
Extraction of Copper from Its Ore 99
Effects of Acid Rain on Plants 103
Wetlands Acid Spill (Acid-base titration) (CBL) 107
How Do Pollutants Affect a Lake? (CBL) 113
Evaluating Fuels (Calorimetry) (CBL) 119

Biological Chemistry

An Introduction to Biological Chemistry 125
Identifying Organic Compounds in Foods 131
Measuring the Release of Energy from Sucrose 135
Diffusion and Cell Membranes 141
Observing Enzyme Detergents **(Inquiry)** 145

Food Chemistry

An Introduction to Food Chemistry 149
Energy Content of Foods (CBL) 155
Buffer Capacity in Commercial Beverages (CBL) 163

Chemical Engineering

An Introduction to Chemical Engineering 169
Micro-Voltaic Cells (CBL) 175
Air Pressure and Piston Design (CBL) 181
Evaporation and Ink Solvents (CBL) 189
A Leaky Reaction (CBL) 197
Solubility and Chemical Fertilizers (CBL) 205

Program Overview

Holt's *Forensics and Applied Science* includes a variety of laboratory investigations as well as introductory sections with assessment. This book can be used as an aid for a semester-long forensic science class, or it can be easily integrated into a biology course, a chemistry course, or another general science program.

This book includes forensics laboratory experiments as well as laboratory experiments in several other areas of applied science. Lab experiments are grouped into categories by procedure type (in the case of forensics labs) and by discipline type (in the case of other applied science labs). Each section of labs begins with a short introduction to ground students in the theory and concepts that are the basis of the experiments they will perform.

TOPIC INTRODUCTIONS

Labs in this book are divided into five applied science topic areas—forensics, environmental chemistry, biological chemistry, food chemistry, and chemical engineering—and each of these groups is prefaced with a **topic introduction.** In a topic introduction, students will learn about an area of applied science by a survey of a contemporary "hot topic"—for example, biomedical engineering—in that area of applied science. Topic introductions also include descriptions of some representative and current careers in the area. A page of topic questions, in which students will be assessed on the material they read in that section, concludes each topic introduction.

PROCEDURE INTRODUCTIONS

Forensic lab experiments are grouped into four categories—density, chromatography, spectroscopy, and identification—based on the type of procedure the experiment uses. Each of these groups of labs begins with a **procedure introduction,** in which students will learn about the science and theory behind the lab technique being used in the labs that follow. Procedure introductions conclude with comprehension questions on the reading, and, if applicable, practice problems in which students practice calculation techniques they will need to apply later.

LAB EXPERIMENTS

The laboratory experiments in this book include detailed instructions to the teacher as well as the student on materials preparation, lab setup, procedure, and tips and tricks for performing the experiment successfully. For some forensics experiments, students will probably need to refer back to the procedure introduction to refresh their memory on how to carry out a certain technique or calculation. In **Inquiry labs,** students will use basic knowledge they have learned about the technique to design their own experimental procedures.

Ordering Lab Materials

Your class and prep time are valuable. Now, it's easier and faster than ever to organize and obtain the materials that you need for all of the labs in the *Forensics and Applied Science* lab program. Either order kits from Science Kit with everything you'll need, or pick and choose with Holt's exclusive Lab Materials QuickList software.

Order kits for one-stop shopping

Holt, Rinehart and Winston has teamed up with Science Kit to offer special kits containing all the materials you'll need for the activities in this lab program. To order, call 1-800-225-5425 or order online at www.hrw.com/catalog.

Lab Materials QuickList

Visit go.hrw.com and type in the keyword **Holt Forensics.** You can customize your list based on which labs you want to do, the number of students and the number of lab groups. A powerful software engine that has been programmed to distinguish between consumable and nonconsumable materials will "do the math." Whether you're examining all of the labs for a whole year or just the labs that you're planning for next week, the software does the hard work of totaling and tallying, telling you *exactly* how much of each material you'll need for the labs and numbers of students, lab groups, and classes that you've selected. Use your materials list to order all of your materials at once. Or use the list to determine what items you need to resupply or supplement your stockroom so that you'll be prepared to do any lab.

After you've created your materials list from the Lab Materials QuickList software, you can use the list to order from Science Kit or Sargent-Welch, or to prepare a purchase order to be sent directly to another scientific materials supplier.

Visit go.hrw.com to learn more about the Lab Materials QuickList software.

Science Kit
1-800-828-7777
www.sciencekit.com

Sargent-Welch
1-800-727-4368
www.sargentwelch.com

1b

Calculator-Based Laboratory Information

Several experiments in Holt's *Forensics and Applied Science* ancillary make use of the Calculator-Based Laboratory 2™ (CBL 2™) data collection interface by Texas Instruments and the Vernier LabPro® data collection interface by Vernier Software & Technology. This lab equipment makes data collection and analysis easy and accurate.

The data collection interface mimics expensive electronic laboratory equipment and allows students to collect experimental data and store it directly onto a graphing calculator. Data are automatically tabulated, and real-time graphs can be displayed. Consequently, students have more time to interpret results. The interface and probes collect experimental data, and the information coming from the probes is automatically recognized and processed through the DataMate™ App that is installed on the calculator.

In various labs, you may use a temperature sensor, a pressure sensor, a voltage sensor, or a colorimeter to collect data. The data can then be analyzed on the calculator to obtain results for the experiment. To perform the calculator-based probeware experiments, you will need

- a graphing calculator (TI-83 Plus and TI-84 Plus)
- the CBL 2™ or LabPro® data collection interface
- the appropriate Texas Instruments or Vernier probe for the experiment
- the probeware experiment

The probeware experiments are available at **go.hrw.com** (keyword **HC6 CBL**). For additional information about the CBL 2™ and LabPro® hardware or software, visit **education.ti.com** or **www.vernier.com.**

Laboratory Safety

In the laboratory, you can engage in hands-on explorations, test your scientific hypotheses, and build practical lab skills. However, while you are working in the lab or in the field, it is your responsibility to protect yourself and your classmates by conducting yourself in a safe manner. You will avoid accidents in the lab by following directions, handling materials carefully, and taking your work seriously. Read the following safety guidelines before working in the lab. Make sure that you understand all safety guidelines before entering the lab.

Before You Begin

- **Read the entire activity before entering the lab.** Be familiar with the instructions before beginning an activity. Do not start an activity until you have asked your teacher to explain any parts of the activity that you do not understand.

- **Student-designed procedures or inquiry activities must be approved by your teacher before you attempt the procedures or activities.**

- **Wear the right clothing for lab work.** Before beginning work, tie back long hair, roll up loose sleeves, and put on any required personal protective equipment as directed by your teacher. Remove your wristwatch and any necklaces or jewelry that could get caught in moving parts. Avoid or confine loose clothing that could knock things over, catch on fire, get caught in moving parts, contact electrical connections, or absorb chemical solutions. Wear pants rather than shorts or skirts. Nylon and polyester fabrics burn and melt more readily than cotton does. Protect your feet from chemical spills and falling objects. Do not wear open-toed shoes, sandals, or canvas shoes in the lab. In addition, chemical fumes may react with and ruin some jewelry, such as pearl jewelry. Do not apply cosmetics in the lab. Some hair care products and nail polish are highly flammable.

- **Do not wear contact lenses in the lab.** Even though you will be wearing safety goggles, chemicals could get between contact lenses and your eyes and could cause irreparable eye damage. If your doctor requires that you wear contact lenses instead of glasses, then you should wear eye-cup safety goggles—similar to goggles worn for underwater swimming—in the lab. Ask your doctor or your teacher how to use eye-cup safety goggles to protect your eyes.

- **Know the location of all safety and emergency equipment used in the lab.** Know proper fire-drill procedures and the location of all fire exits. Ask your teacher where the nearest eyewash stations, safety blankets, safety shower, fire extinguisher, first-aid kit, and chemical spill kit are located. Be sure that you know how to operate the equipment safely.

Name _____ Class _____ Date _____

Lab Safety *continued*

While You Are Working

- **Always wear a lab apron and safety goggles.** Wear these items even if you are not working on an activity. Labs contain chemicals that can damage your clothing, skin, and eyes. If your safety goggles cloud up or are uncomfortable, ask your teacher for help. Lengthening the strap slightly, washing the goggles with soap and warm water, or using an anti-fog spray may help the problem.

- **NEVER work alone in the lab.** Work in the lab only when supervised by your teacher. Do not leave equipment unattended while it is in operation.

- **Perform only activities specifically assigned by your teacher.** Do not attempt any procedure without your teacher's direction. Use only materials and equipment listed in the activity or authorized by your teacher. Steps in a procedure should be performed only as described in the activity or as approved by your teacher.

- **Keep your work area neat and uncluttered.** Have only books and other materials that are needed to conduct the activity in the lab. Keep backpacks, purses, and other items in your desk, locker, or other designated storage areas.

- **Always heed safety symbols and cautions listed in activities, listed on handouts, posted in the room, provided on chemical labels, and given verbally by your teacher.** Be aware of the potential hazards of the required materials and procedures, and follow all precautions indicated.

- **Be alert, and walk with care in the lab.** Be aware of others near you and your equipment.

- **Do not take food, drinks, chewing gum, or tobacco products into the lab.** Do not store or eat food in the lab.

- **NEVER taste chemicals or allow them to contact your skin.** Keep your hands away from your face and mouth, even if you are wearing gloves.

- **Exercise caution when working with electrical equipment.** Do not use electrical equipment with frayed or twisted wires. Be sure that your hands are dry before using electrical equipment. Do not let electrical cords dangle from work stations. Dangling cords can cause you to trip and can cause an electrical shock. The area under and around electrical equipment should be dry; cords should not lie in puddles of spilled liquid.

- **Use extreme caution when working with hot plates and other heating devices.** Keep your head, hands, hair, and clothing away from the flame or heating area. Remember that metal surfaces connected to the heated area will become hot by conduction. Gas burners should be lit only with a spark lighter, not with matches. Make sure that all heating devices and gas valves are turned off before you leave the lab. Never leave a heating device unattended when it is in use. Metal, ceramic, and glass items do not necessarily look hot when they are hot. Allow all items to cool before storing them.

- **Do not fool around in the lab.** Take your lab work seriously, and behave appropriately in the lab. Lab equipment and apparatus are not toys; never use lab time or equipment for anything other than the intended purpose. Be aware of the safety of your classmates as well as your safety at all times.

Emergency Procedures

- **Follow standard fire-safety procedures.** If your clothing catches on fire, do not run; WALK to the safety shower, stand under it, and turn it on. While doing so, call to your teacher. In case of fire, alert your teacher and leave the lab.

- **Report any accident, incident, or hazard—no matter how trivial—to your teacher immediately.** Any incident involving bleeding, burns, fainting, nausea, dizziness, chemical exposure, or ingestion should also be reported immediately to the school nurse or to a physician. If you have a close call, tell your teacher so that you and your teacher can find a way to prevent it from happening again.

- **Report all spills to your teacher immediately.** Call your teacher rather than trying to clean a spill yourself. Your teacher will tell you whether it is safe for you to clean up the spill; if it is not safe, your teacher will know how to clean up the spill.

- **If you spill a chemical on your skin, wash the chemical off in the sink and call your teacher.** If you spill a solid chemical onto your clothing, brush it off carefully without scattering it onto somebody else and call your teacher. If you get liquid on your clothing, wash it off right away by using the faucet at the sink and call your teacher. If the spill is on your pants or something else that will not fit under the sink faucet, use the safety shower. Remove the pants or other affected clothing while you are under the shower, and call your teacher. (It may be temporarily embarrassing to remove pants or other clothing in front of your classmates, but failure to flush the chemical off your skin could cause permanent damage.)

- **If you get a chemical in your eyes, walk immediately to the eyewash station, turn it on, and lower your head so your eyes are in the running water.** Hold your eyelids open with your thumbs and fingers, and roll your eyeballs around. You have to flush your eyes continuously for at least 15 minutes. Call your teacher while you are doing this.

When You Are Finished

- **Clean your work area at the conclusion of each lab period as directed by your teacher.** Broken glass, chemicals, and other waste products should be disposed of in separate, special containers. Dispose of waste materials as directed by your teacher. Put away all material and equipment according to your teacher's instructions. Report any damaged or missing equipment or materials to your teacher.

Name _____ Class _____ Date _____

Lab Safety *continued*

- **Wash your hands with soap and hot water after each lab period.** To avoid contamination, wash your hands at the conclusion of each lab period, and before you leave the lab.

Safety Symbols

Before you begin working in the lab, familiarize yourself with the following safety symbols, which are used throughout your textbook, and the guidelines that you should follow when you see these symbols.

 EYE PROTECTION

- **Wear approved safety goggles as directed.** Safety goggles should be worn in the lab at all times, especially when you are working with a chemical or solution, a heat source, or a mechanical device.

- **If chemicals get into your eyes, flush your eyes immediately.** Go to an eyewash station immediately, and flush your eyes (including under the eyelids) with running water for at least 15 minutes. Use your thumb and fingers to hold your eyelids open and roll your eyeball around. While doing so, ask another student to notify your teacher.

- **Do not wear contact lenses in the lab.** Chemicals can be drawn up under a contact lens and into the eye. If you must wear contacts prescribed by a physician, tell your teacher. In this case, you must also wear approved eye-cup safety goggles to help protect your eyes.

- **Do not look directly at the sun or any light source through any optical device or lens system, and do not reflect direct sunlight to illuminate a microscope.** Such actions concentrate light rays to an intensity that can severely burn your retinas, which may cause blindness.

 CLOTHING PROTECTION

- **Wear an apron or lab coat at all times in the lab to prevent chemicals or chemical solutions from contacting skin or clothes.**

- **Tie back long hair, secure loose clothing, and remove loose jewelry so that they do not knock over equipment, get caught in moving parts, or come into contact with hazardous materials.**

| Lab Safety *continued*

 HYGIENIC CARE

- **Keep your hands away from your face and mouth while you are working in the lab.**
- **Wash your hands thoroughly before you leave the lab.**
- **Remove contaminated clothing immediately.** If you spill caustic substances on your skin or clothing, use the safety shower or a faucet to rinse. Remove affected clothing while you are under the shower, and call to your teacher. (It may be temporarily embarrassing to remove clothing in front of your class-mates, but failure to rinse a chemical off your skin could result in permanent damage.)
- **Launder contaminated clothing separately.**
- **Use the proper technique demonstrated by your teacher when you are handling bacteria or other microorganisms.** Treat all microorganisms as if they are pathogens. Do not open Petri dishes to observe or count bacterial colonies.
- **Return all stock and experimental cultures to your teacher for proper disposal.**

 SHARP-OBJECT SAFETY

- **Use extreme care when handling all sharp and pointed instruments, such as scalpels, sharp probes, and knives.**
- **Do not cut an object while holding the object in your hand.** Cut objects on a suitable work surface. Always cut in a direction away from your body.
- **Do not use double-edged razor blades in the lab.**

 GLASSWARE SAFETY

- **Inspect glassware before use; do not use chipped or cracked glassware.** Use heat-resistant glassware for heating materials or storing hot liquids, and use tongs or a hot mitt to handle this equipment.
- **Do not attempt to insert glass tubing into a rubber stopper without specific instructions from your teacher.**
- **Notify immediately your teacher if a piece of glassware or a light bulb breaks.** Do not attempt to clean up broken glass unless your teacher directs you to do so.

☣ PROPER WASTE DISPOSAL

- **Clean and sanitize all work surfaces and personal protective equipment after each lab period as directed by your teacher.**

- **Dispose of all sharp objects (such as broken glass) and other contaminated materials (biological or chemical) in special containers only as directed by your teacher.** Never put these materials into a regular waste container or down the drain.

⚡ ELECTRICAL SAFETY

- **Do not use equipment with frayed electrical cords or loose plugs.**

- **Fasten electrical cords to work surfaces by using tape.** Doing so will prevent tripping and will ensure that equipment will not fall off the table.

- **Do not use electrical equipment near water or when your clothing or hands are wet.**

- **Hold the rubber cord when you plug in or unplug equipment.** Do not touch the metal prongs of the plug, and do not unplug equipment by pulling on the cord.

- **Wire coils on hot plates may heat up rapidly.** If heating occurs, open the switch immediately and use a hot mitt to handle the equipment.

🔥 HEATING SAFETY

- **Be aware of any source of flames, sparks, or heat (such as open flames, electric heating coils, or hot plates) before working with flammable liquids or gases.**

- **Avoid using open flames.** If possible, work only with hot plates that have an on/off switch and an indicator light. Do not leave hot plates unattended. Do not use alcohol lamps. Turn off hot plates and open flames when they are not in use.

- **Never leave a hot plate unattended while it is turned on or while it is cooling off.**

- **Know the location of lab fire extinguishers and fire-safety blankets.**

- **Use tongs or appropriate insulated holders when handling heated objects.** Heated objects often do not appear to be hot. Do not pick up an object with your hand if it could be warm.

- **Keep flammable substances away from heat, flames, and other ignition sources.**

- **Allow all equipment to cool before storing it.**

Lab Safety *continued*

 FIRE SAFETY

- **Know the location of lab fire extinguishers and fire-safety blankets.**

- **Know your school's fire-evacuation routes.**

- **If your clothing catches on fire, walk (do not run) to the emergency lab shower to put out the fire. If the shower is not working, STOP, DROP, and ROLL!**

 **CAUSTIC SUBSTANCES**

- **If a chemical gets on your skin, on your clothing, or in your eyes, rinse it immediately and alert your teacher.**

- **If a chemical is spilled on the floor or lab bench, alert your teacher, but do not clean it up yourself unless your teacher directs you to do so.**

 CHEMICAL SAFETY

- **Always wear safety goggles, gloves, and a lab apron or coat to protect your eyes and skin when you are working with any chemical or chemical solution.**

- **Do not taste, touch, or smell any chemicals or bring them close to your eyes unless specifically instructed to do so by your teacher.** If your teacher tells you to note the odor of a substance, do so by waving the fumes toward you with your hand. Do not pipette any chemicals by mouth; use a suction bulb as directed by your teacher.

- **Know where the emergency lab shower and eyewash stations are and how to use them.** If you get a chemical on your skin or clothing, wash it off at the sink while calling to your teacher.

- **Always handle chemicals or chemical solutions with care.** Check the labels on bottles, and observe safety procedures. Label beakers and test tubes containing chemicals.

- **For all chemicals, take only what you need.** Do not return unused chemicals or solutions to their original containers. Return unused reagent bottles or containers to your teacher.

- **NEVER take any chemicals out of the lab.**

- **Do not mix any chemicals unless specifically instructed to do so by your teacher.** Otherwise harmless chemicals can be poisonous or explosive if combined.

- **Do not pour water into a strong acid or base.** The mixture can produce heat and can splatter.

- **Report all spills to your teacher immediately.** Spills should be cleaned up promptly as directed by your teacher.

Laboratory Techniques

FIGURE A

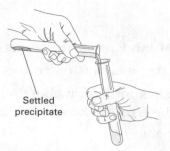

Settled
precipitate

FIGURE B

FIGURE C

HOW TO DECANT AND TRANSFER LIQUIDS

1. The safest way to transfer a liquid from a graduated cylinder to a test tube is shown in Figure A. The liquid is transferred at arm's length, with the elbows slightly bent. This position enables you to see what you are doing while maintaining steady control of the equipment.

2. Sometimes, liquids contain particles of insoluble solids that sink to the bottom of a test tube or beaker. Use one of the methods shown above to separate a supernatant (the clear fluid) from insoluble solids.

a. Figure B shows the proper method of decanting a supernatant liquid from a test tube.

b. Figure C shows the proper method of decanting a supernatant liquid from a beaker by using a stirring rod. The rod should touch the wall of the receiving container. Hold the stirring rod against the lip of the beaker containing the supernatant. As you pour, the liquid will run down the rod and fall into the beaker resting below. When you use this method, the liquid will not run down the side of the beaker from which you are pouring.

HOW TO HEAT SUBSTANCES AND EVAPORATE SOLUTIONS

1. Use care in selecting glassware for high-temperature heating. The glassware should be heat resistant.

2. When heating glassware by using a gas flame, use a ceramic-centered wire gauze to protect glassware from direct contact with the flame. Wire gauzes can withstand extremely high temperatures and will help prevent glassware from breaking. Figure D shows the proper setup for evaporating a solution over a water bath.

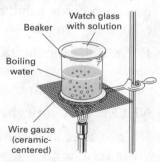

Beaker

Watch glass
with solution

Boiling
water

Wire gauze
(ceramic-
centered)

FIGURE D

3. In some experiments, you are required to heat a substance to high temperatures in a porcelain crucible. Figure E shows the proper apparatus setup used to accomplish this task.

4. Figure F shows the proper setup for evaporating a solution in a porcelain evaporating dish with a watch glass cover that prevents spattering.

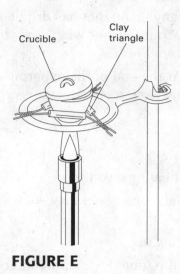

Crucible Clay triangle

FIGURE E

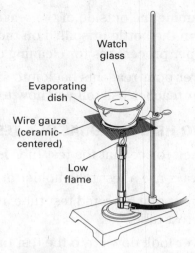

Watch glass

Evaporating dish

Wire gauze (ceramic-centered)

Low flame

FIGURE F

5. Glassware, porcelain, and iron rings that have been heated may *look* cool after they are removed from a heat source, but these items can still burn your skin even after several minutes of cooling. Use tongs, test-tube holders, or heat-resistant mitts and pads whenever you handle these pieces of apparatus.

6. You can test the temperature of beakers, ring stands, wire gauzes, or other pieces of apparatus that have been heated by holding the back of your hand close to their surfaces before grasping them. You will be able to feel any energy as heat generated from the hot surfaces. DO NOT TOUCH THE APPARATUS. Allow plenty of time for the apparatus to cool before handling.

HOW TO POUR LIQUID FROM A REAGENT BOTTLE

1. Read the label at least three times before using the contents of a reagent bottle.

2. Never lay the stopper of a reagent bottle on the lab table.

3. When pouring a caustic or corrosive liquid into a beaker, use a stirring rod to avoid drips and spills. Hold the stirring rod against the lip of the reagent bottle. Estimate the amount of liquid you need, and pour this amount along the rod, into the beaker. See Figure G.

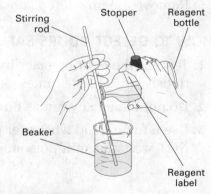

Stirring rod Stopper Reagent bottle

Beaker

Reagent label

FIGURE G

4. Extra precaution should be taken when handling a bottle of acid. Remember the following important rules: Never add water to any concentrated acid, particularly sulfuric acid, because the mixture can splash and will generate a lot of energy as heat. To dilute any acid, add the acid to water in small quantities while stirring slowly. Remember the "triple A's"—*Always Add Acid* to water.

5. Examine the outside of the reagent bottle for any liquid that has dripped down the bottle or spilled on the counter top. Your teacher will show you the proper procedures for cleaning up a chemical spill.

6. Never pour reagents back into stock bottles. At the end of the experiment, your teacher will tell you how to dispose of any excess chemicals.

HOW TO HEAT MATERIAL IN A TEST TUBE

1. Check to see that the test tube is heat resistant.

2. Always use a test tube holder or clamp when heating a test tube.

3. Never point a heated test tube at anyone, because the liquid may splash out of the test tube.

4. Never look down into the test tube while heating it.

5. Heat the test tube from the upper portions of the tube downward, and continuously move the test tube, as shown in Figure H. Do not heat any one spot on the test tube. Otherwise, a pressure buildup may cause the bottom of the tube to blow out.

HOW TO USE A MORTAR AND PESTLE

1. A mortar and pestle should be used for grinding only one substance at a time. See Figure I.

2. Never use a mortar and pestle for simultaneously mixing different substances.

3. Place the substance to be broken up into the mortar.

4. Pound the substance with the pestle, and grind to pulverize.

5. Remove the powdered substance with a porcelain spoon.

HOW TO DETECT ODORS SAFELY

1. Test for the odor of gases by wafting your hand over the test tube and cautiously sniffing the fumes as shown in Figure J.

2. Do not inhale any fumes directly.

3. Use a fume hood whenever poisonous or irritating fumes are present. DO NOT waft and sniff poisonous or irritating fumes.

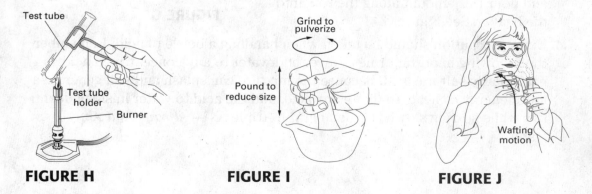

FIGURE H   **FIGURE I**   **FIGURE J**

Name _____ Class _____ Date _____

Topic Introduction

An Introduction to Forensic Science

When most people think about forensic science, they recall crime dramas from television, in which a crime scene investigator seizes the smallest bits of evidence and ties them together to reconstruct the crime, enabling the investigator to identify even the most cunning criminal. For example, the evidence associated with a crime may be samples of hair, paint, glass, soil, blood, or plant material. A forensic scientist tests these pieces of evidence using various analytical procedures such as density tests, color tests, chromatography, and DNA tests.

Many of the scientists who work in forensics are involved in law enforcement work. But there's much more to forensics than that. Basically, **forensics** *is making knowledge and information available in a public forum, such as a court of law.* A *forensic scientist* is a person who applies scientific knowledge and techniques to the investigation of evidence for the purpose of identification.

Sometimes, a single scientific technique can be applied to solve many different scientific problems as well as answer forensics questions. One example is the rapidly developing field of mitochondrial DNA analysis.

A Current Hot Topic: Forensic mtDNA Analysis

All cells contain mitochondria, shown below in **Figure 1,** which are small structures about the size of a bacterium. Mitochondria act as "power generators" for cells, making ATP. They are found in the cell material surrounding the nucleus, numbering from several to a thousand or more. They have their own DNA, and each mitochondrion contains several copies of its DNA.

FIGURE 1 A MITOCHONDRION WITH ITS DNA

Mitochondria contain their own DNA, referred to as mtDNA.

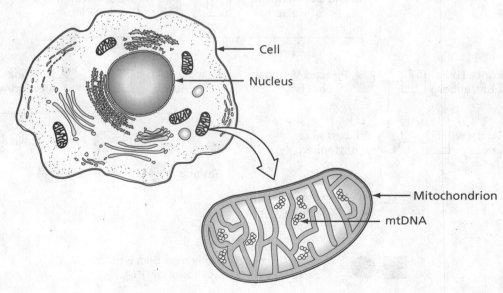

An Introduction to Forensic Science *continued*

A person's mitochondria and mitochondrial DNA are always inherited from the mother, whereas nuclear DNA is inherited partly from the mother and partly from the father. One advantage of using mtDNA for forensic analysis is the fact that there are great numbers of mitochondria in each cell. This abundance allows a very small sample of material to provide a large mtDNA sample. Another advantage is that mtDNA exists in all types of cells and therefore any type of body tissue can be used, including bone.

FORENSIC GENEALOGY

The very new technique of mtDNA analysis can aid in the identification of human remains, even solving mysteries that are nearly a century old.

In 1918, during the Russian Revolution, Czar Nicholas II, his wife Czarina Alexandra, and their children were executed. Many legends and rumors surrounded their deaths and the location of their grave. Years later, a woman in Paris claimed to be their daughter Anastasia, who had somehow escaped.

Investigators obtained mtDNA from a grave site rumored to be the place where the bodies of the czar's family were buried for the purpose of *forensic genealogy.* Using forensic genealogy, one attempts to establish family relationships between the living and the deceased. In this case, Prince Philip of Greece, the husband of Great Britain's Queen Elizabeth II, agreed to provide an mtDNA sample. Because he is a direct maternal descendant of the czarina's mother, he has the same mtDNA pattern as the czarina. His pattern matched those from the gravesite, proving that the site was the final resting place of the czar's family. A similar analysis disproved the claim of the woman from Paris. A diagram of the maternal lineage that connected these relatives is shown below in **Figure 2.**

FIGURE 2 PARTIAL GENEALOGY OF EUROPEAN ROYAL FAMILIES

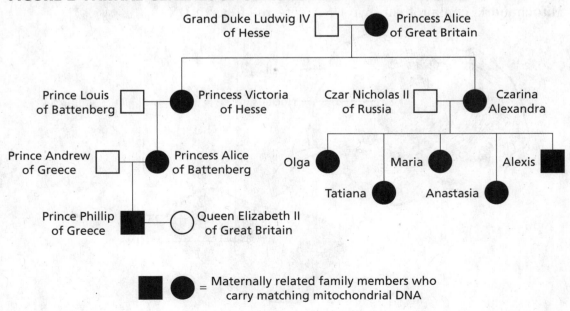

Name _____ Class _____ Date _____

An Introduction to Forensic Science *continued*

MOLECULAR ANTHROPOLOGY

Mitochondrial DNA analysis is also being used to probe even older mysteries—those of human origins. Until recently, these studies have been done only on the physical characteristics of humans and their ancestors. In *molecular anthropology*, however, human characteristics can be studied genetically. A molecular anthropologist studies the similarities and differences of DNA and mtDNA of different groups of living humans to determine evolutionary and migratory patterns. Molecular anthropologists also study the mtDNA of modern humans and that of fossilized human remains in an effort to determine the early origins of the human race. Molecular anthropologists try to determine when and where the first humans lived and how and where those early humans migrated.

DISEASE AND MITOCHONDRIAL GENETICS

There are also applications of mtDNA analysis in medicine. To date, forensic investigations have found mutations in the genetic code of mtDNA that are related to disease. Some of these mutations may cause more than one disease or may cause different diseases to occur at different periods in a person's lifetime. Other investigations have found that certain toxins and drugs can cause mutations in mtDNA or interfere with its replication. One such relationship has been found with some of the drugs used to treat HIV/AIDS. The amount of mtDNA in these patients is smaller than normal, and these patients develop severe muscle weakness as a result.

Aging is the most common of all genetic diseases. Forensic investigators are trying to determine how the inherited genetics of mtDNA and any subsequent mutations affect the aging process. Current investigations into Parkinson's disease and Alzheimer's disease are attempting to determine if these diseases are related to mtDNA mutations.

Careers in Forensic Science

Mitochondrial DNA analysis is but one of many new technologies being developed for use in forensic science. With the rapid growth in this field, there are many different career opportunities in forensic science.

QUESTIONED DOCUMENT EXAMINATION

Forensic scientists who work in the area of questioned document examination determine who is the author of a document or how a document was created. Specifically, these scientists may try to determine whose handwriting is on the document, the machine (typewriter, copier, or fax) and the inks used to create the document, and the material on which the document is written. Researchers are now looking at ways to analyze the language patterns in a document, including particular words and phrases, the sentence construction, and verb tenses, to help identify the author of the document.

LATENT FINGERPRINTS RESEARCH

Many forensic analysts try to find fingerprints at a crime scene: this evidence is vital to many investigations. The old method of "dusting" for fingerprints is very time consuming, and some prints may be missed. New methods are being developed that allow an entire room, or even an entire house, to be scanned for fingerprints in a matter of hours. One of the new methods uses fumes, produced when dried "super glue" is burned, to make *latent*, or hidden, fingerprints easily visible so that they can be analyzed. Researchers in this area have to understand how particular chemicals react with the chemicals in a fingerprint.

TRACE EVIDENCE EXAMINATION

Glass fragments, paint chips, and gunpowder residue are examples of trace evidence. A forensic scientist in this area analyzes these very small samples by using both physical and chemical tests, which may destroy the sample. New research is underway to find ways to analyze very small samples. Some methods being investigated are sophisticated types of chromatography, laser scanning, and fluorescent imaging.

FIREARMS ANALYSIS

A gun's barrel leaves a set of scratches on the bullet fired from the gun. Because these scratches are unique to each gun, it is possible to identify the particular gun that fired a bullet. The firearms analyst compares these scratches to other scratched bullets in an attempt to match the bullet to a particular gun. Firearms analysts hope to create a computer database filled with gun scratch patterns. Then, the scratch pattern on a bullet fired from a gun would be scanned into a computer for comparison with the database of scratch patterns from specific weapons.

FORENSIC DENTISTRY

If a dead body has been badly burned or has decomposed, identification of the remains may be very difficult. In such cases, the only parts of the body left intact are often the jaws and teeth. A forensic dentist can identify human remains based on dental records, if the records exist. The forensic dentist may perform a dental exam during an autopsy, which may include X rays and charts of the teeth and the skull. In other cases, the forensic dentist may examine bite marks on a victim for comparison with the tooth patterns of a suspect. The examination of dental injuries that occurred during a crime is another task that may be required of the forensic dentist.

An Introduction to Forensic Science *continued*

Topic Questions

1. Describe in one sentence what a forensic scientist does.

Answers may vary. Sample answer: A forensic scientist uses scientific techniques to identify pieces of evidence and make the results known to the public.

2. Where is mtDNA found, and how does it differ from the DNA in the nucleus?

mtDNA is found in mitochondria, which are small structures in the material surrounding the nucleus of a cell. Mitochondrial DNA has a very small number of base pairs compared with the DNA in the nucleus. mtDNA is inherited from only the mother, whereas nuclear DNA is inherited from both parents.

3. What new method is being developed for determining the author of a document?

analysis of the various aspects—such as the words and sentence structure—of the language used to write the document

4. How can a firearms analyst tell what gun a certain bullet was fired from?

Each gun leaves a unique pattern of scratches on a bullet fired from that gun. Analysis of the scratch pattern on a bullet plus comparison with a database of known firearms can identify the weapon that fired the bullet.

5. A corpse is discovered buried in a field. It may be a person who was reported missing a few years ago and is now feared dead. The body is decomposed and is not easily identifiable. Suggest a way that the body might be identified.

Answers may vary. Students' answers may mention DNA or mtDNA analysis in which the DNA of the deceased is compared with that of living relatives of the missing person. Forensic dentistry, matching the dental records of the missing person with the teeth and jaw of the body, is another option.

Procedure Introduction

A Lesson on Density

You come out of the grocery store and find that someone has broken the brake light on your car. You find a small shard of glass that does not match your brake light but appears to match a broken headlight on the car in the next space. You get a loose piece of headlight glass and the license number from the other car. Now, how could you determine whether the two pieces of glass come from the same headlight? Density may be the key.

BACKGROUND

Density *is the ratio of the mass of an object to the volume of the object.* This ratio can be expressed as an equation, as shown below, where d is density, m is mass, and V is volume.

$$d = \frac{m}{V}$$

When an object is placed in fluid, the difference between its density and the fluid's density determines whether the object will sink or float. One of three things will happen when an object is placed in fluid:

- If the density of the object is greater than the density of the fluid, the object will sink in the fluid, and the volume of fluid *displaced* (meaning the amount the fluid level rises when the object is dropped into it) will be the same as the volume of the object.

- If the density of the object is less than the density of the fluid, the object will float on the surface of the fluid and the volume of fluid displaced will have a weight equal to the weight of the object.

- When the density of the object is the same as the density of the fluid, the object will neither sink nor float: it will remain suspended in the fluid, and the volume of fluid displaced will be equal to the volume of the object.

An object made of a certain material will have a density that is characteristic of that material. Therefore, density can be used to identify the material an object is made of. This has been known since ancient times: Archimedes, a Greek who lived in the third century BCE, discovered an important fact, now known as Archimedes' principle, about buoyant force. **Archimedes' principle** *states that when an object is placed in a fluid, a buoyant force is exerted by the fluid on the object that is equal to the weight of the fluid displaced by the object.* Archimedes' principle can be stated in a very simple equation, shown below.

buoyant force = weight of displaced fluid

A Lesson on Density *continued*

Because of the buoyant force, an object will weigh less submerged than it weighs in air. A floating object, on the other hand, appears to be weightless. This is because a floating object is less dense than the fluid that it floats in, so only part of its volume displaces the fluid. The weight of the displaced fluid is the same as the weight of the entire object. Archimedes' principle can be used in various interesting and creative ways, depending on the requirements of a particular sample, to determine the density of an object.

How to Determine Density
DENSITY DETERMINATION OF AN IRREGULARLY SHAPED OBJECT

The density of a relatively large but irregularly shaped object is most easily determined by using Archimedes' principle. In this method, the weight, or apparent mass, of an object is compared with its apparent mass when it is submerged in a fluid. The equation for determining the density of an object submerged in a fluid by this method is shown below.

$$density_{object} = \frac{apparent\ mass_{object\ in\ air}}{\left(\dfrac{apparent\ mass_{object\ in\ air} - apparent\ mass_{object\ in\ fluid}}{density_{fluid}}\right)}$$

This equation can be used to determine the density of an object when it is difficult or impossible to accurately find the volume of the object. A piece of shattered glass with sharp edges and points is an example of the type of object for which this method of density determination is suitable.

The advantage of this method, as you can see by glancing at the equation, is that you don't have to measure volumes at all: you need to know only the density of the fluid used and the apparent mass of the object in air and in the fluid. Water is often used as the fluid for this technique because its density of about 1.00 g/cm³ simplifies calculations, but any fluid can be used provided that it has a density that is less than the density of the object.

In the equation shown above, the apparent mass difference divided by the density of the fluid yields the volume of the object. The equation below shows how the units in the equation above cancel to arrive at the same simple equation shown on the previous page, which is mass per unit volume.

$$density_{object} = \frac{apparent\ mass_{object\ in\ air}}{\left(\dfrac{apparent\ mass_{object\ in\ air} - apparent\ mass_{object\ in\ fluid}}{density_{fluid}}\right)}$$

$$= \frac{m}{\left(\dfrac{m - m}{\dfrac{m}{V}}\right)} = \frac{m}{\left(\dfrac{m}{\dfrac{m}{V}}\right)} = \frac{m}{V}$$

DENSITY DETERMINATION OF A VERY SMALL OBJECT

To determine the density of a very small object such as a tiny piece of glass or plastic, it is often more convenient and accurate to place the object in a fluid of known density to determine whether the object sinks or floats. The density of the fluid can then be adjusted by adding another fluid, one that is miscible with the first fluid but has a different density, dropwise until the object being tested neither floats nor sinks. As described earlier, when this condition is reached, the density of the fluid and the density of the object are the same.

Accurate results require careful recording of the number of drops of fluid added, so you can calculate an accurate final density. Knowing the exact number of drops of each fluid added and the density of each fluid, you can calculate the final volume of the mixtures by using *weighted averages.* In using weighted averages for these calculations, a "weight" is given to the density of each fluid based on how much of each was present in the mixture at the point where it achieved the same density as the glass sample. This simply means multiplying the density of each fluid by the number of drops that was present. Then you add the two weighted densities together and divide that by the total number of drops of fluid in the mixture: this weighted average will be the final density of the mixture.

For example, say you have glass sample that is suspended in a mixture. To make the mixture, you had added 30 drops of a fluid with a density of 2.50 g/mL, plus 5 drops of water, which has a density of about 0.998 g/mL. First, you "weight" each density by the number of drops, and add them together:

$$(30 \text{ drops} \times 2.50 \text{ g/mL}) + (5 \text{ drops} \times 0.998 \text{ g/mL}) = 80.0 \text{ drops} \bullet \text{g/mL}$$

To get the weighted average, divide this weighted total by the total number of drops:

$$\text{density} = \frac{80.0 \text{ drops} \bullet \text{g/mL}}{35 \text{ drops}} = 2.29 \text{ g/mL}$$

Because 2.29 g/mL is the final density of the mixture when the glass sample is suspended, 2.29 g/mL is also the density of the glass sample.

Name _____ Class _____ Date _____

A Lesson on Density *continued*

COMPARING DENSITIES OF TWO OBJECTS BY USING A COLUMN

Sometimes it is less important to find the density of an object than to simply find out if its density is the same as that of another object. Comparing the density of two objects can be performed in a single step by the use of a density gradient column. The column contains several fluids of progressively greater density. Dropping the objects into the column and allowing them to come to rest is all that is required for the comparison. If the objects come to rest at the same position in the column, you know that they have the same density. Otherwise, you know the densities are different. **Figure 1** below shows a diagram of such a column containing various materials that have settled to the layers that indicate their densities (shown in grams per milliliter).

FIGURE 1 DENSITY-LAYERS COLUMN

Objects having different densities come to rest at different levels in a density column. Each object placed in the density column sinks until it reaches a point where its density is less than the density of the surrounding fluid. Notice that each object floats on the fluid that has a density greater than the density of the object.

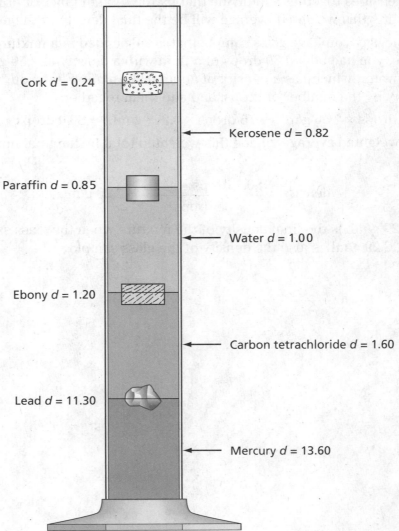

Cork *d* = 0.24

Kerosene *d* = 0.82

Paraffin *d* = 0.85

Water *d* = 1.00

Ebony *d* = 1.20

Carbon tetrachloride *d* = 1.60

Lead *d* = 11.30

Mercury *d* = 13.60

Topic Questions

1. What is the mathematical definition of density?

mass divided by volume

2. State Archimedes' principle.

The buoyant force on an object in a fluid is equal to the weight of the fluid

displaced by the object.

3. An object placed in a fluid sinks and displaces a volume of fluid. What is the relationship between the volume of fluid displaced and the volume of the object?

The volume of fluid displaced is the same as the volume of the object.

4. If an object is submerged in a fluid, how is the weight of the object affected? Why?

The weight of the object decreases. The fluid exerts a force on the object

equal to the weight of the fluid displaced by the object.

5. Two pieces of plastic are placed in a density column. Piece A comes to rest close to the middle of the column, whereas piece B sinks to the bottom. What could you say about the densities of the two pieces?

Piece A has a density about equal to the density of the middle part of the

column. Because piece B sinks, it has a density that is greater than piece A

and greater than the densest part of the column.

6. Suppose that to determine the density of an object, you use the weight-difference technique. What information is necessary for the determination to be successful? Explain.

The density of the fluid used. The density of the fluid is used in the calcula-

tion to determine the volume of the object. The volume is required to deter-

mine the density because density is mass per unit volume.

Name _____ Class _____ Date _____

A Lesson on Density *continued*

Practice Problems

1. When a bracelet with an apparent mass of 383 g in air is submerged in water, a scale measures an apparent mass of 349 g. What is the density of the bracelet? Is the bracelet made of pure gold or something else? Explain. (Use 1.00 g/cm³ as the density of water and 19.3 g/cm³ as the density of pure gold.)

The density of the bracelet is 11.3 g/cm³. The bracelet is not made of pure gold because the density of the bracelet is less than the density of pure gold. The bracelet could be made from an alloy containing gold.

Calculation:

$$density_{object} = \frac{apparent\ mass_{object\ in\ air}}{\left(\dfrac{apparent\ mass_{object\ in\ air} - apparent\ mass_{object\ in\ water}}{density_{water}}\right)}$$

$$= \frac{383\ g}{\left(\dfrac{383\ g - 349\ g}{1.00\ g/cm^3}\right)} = \frac{383\ g}{34.0\ cm^3} = 11.3\ g/cm^3$$

2. In order to determine the density of a small shard of glass so you can identify its origin, you are using the method of weighted averages of densities of two fluids. After placing the sample in a test tube, you carefully measure 35 drops of a fluid with a density of 3.40 g/mL into the test tube. After adding 9 drops of water ($d = 0.998$ g/mL) into the test tube, the glass sample is suspended, neither sinking nor floating. What is the sample's density?

The density of the glass piece is 2.91 g/mL.

Calculation:

$$density = \frac{(35\ drops \times 3.40\ g/mL) + (9\ drops \times 0.998\ g/mL)}{44\ drops}$$

$$= \frac{119\ drops \cdot g/mL + 8.98\ drops \cdot g/mL}{44\ drops} = \frac{128\ drops \cdot g/mL}{44\ drops} = 2.91\ g/mL$$

The Hit and Run (density determination using Archimedes' principle)

Teacher Notes

TIME REQUIRED one 50-minute class period

LAB RATINGS

Easy ← 1   2   3   4 → Hard

Teacher Preparation–1
Student Setup–1
Concept Level–3
Cleanup–1

SKILLS ACQUIRED

Designing experiments
Experimenting
Collecting data
Inferring
Interpreting
Measuring
Organizing and analyzing data
Communicating

SCIENTIFIC METHODS

Make Observations Students will make observations and collect data in order to answer the experimental question.

Analyze the Results Students will analyze their data in a systematic fashion.

Draw Conclusions Students will draw conclusions from their data in order to determine the answer to the experimental question.

Communicate the Results Students will clearly communicate their conclusions based on the outcome of the experiment.

MATERIALS (PER LAB GROUP)

- balance, mechanical with support platforms
- beaker, 250 mL or 400 mL
- glass fragment (fragment should be large enough to fit into the beaker without touching the sides)
- scissors
- tweezers
- water

SAFETY CAUTIONS

No chemicals are used in this activity. However, because glass fragments are manipulated, there is the potential for students to be cut or punctured by glass. Students should wear appropriate protective gloves and work with caution when handling glass fragments (cloth gloves will best protect from cuts and punctures). As usual, appropriate protective goggles and aprons should be worn in the event an accident may produce flying projectiles.

Students should use tweezers to handle the glass sample, to prevent cuts.

DISPOSAL

Notify janitorial staff when disposing of broken glass. The glass samples should be stored for future use.

NOTES ON TECHNIQUE

The usual method of determining density by dividing the mass of an object by its volume is not possible in this instance because of the irregular shape of the glass fragment to be tested. Consequently, the use of Archimedes' principle, in which the volume of an irregular object is calculated by measuring the volume of water displaced by the object when submerged, is the only practical alternative. Students seldom use the platform support feature on mechanical balances, so you may have to demonstrate its use. Weighing the glass sample in air should be no problem, but you may need to demonstrate weighing the sample in a beaker filled with water.

TIPS AND TRICKS

As a discussion question, you might want to ask why a large fish is best left in the water and netted rather than being lifted into a boat.

Because the individual glass samples will vary in size, each lab group's results will differ. Make sure to use glass of different types for different lab groups or classes so that the interpretations will vary.

SAMPLE DATA AND CALCULATION

For a sample of headlight glass:

mass in air = 29.88 g

mass when immersed in water (calculated by subtracting [beaker + water mass] from [beaker + glass mass]) = 17.91 g

water volume change = 29.88 g – 17.91 g = 11.97 mL

density = 29.88 g/11.97 mL = 2.50 g/mL

Again, varying the glass type from one group to another forces the students to concentrate more on their own work rather than rely on others.

The Hit and Run

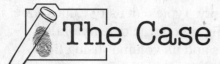

The Case

Deputy Ramirez says it could have been worse. Even so, Shanna's compact car was totaled. It lay steaming in the ditch alongside the road, with its side smashed in where the other car had rammed it. The fire department crew had to pry the doors open to get Shanna out. Surprisingly, she'd escaped with a few broken bones and some cuts and bruises, but nothing worse.

Before the ambulance rolled away, Shanna told Deputy Ramirez that all she could remember was a green sedan rushing at her. Although there weren't any other witnesses at this lonely country intersection, Deputy Ramirez swiftly began piecing together how the crash happened.

He knew Shanna had been driving on the state highway, which has no stop sign. The other car had failed to heed its stop sign, though. After the impact, the other car backed up and peeled off down the road, judging by the skid marks. As Deputy Ramirez strode back to where the impact occurred, he knelt down and picked up a sizable glass fragment and placed it in an evidence bag.

"This glass fragment most likely came from that wreck. I know it could not have come from a much earlier crash," Deputy Ramirez tells you. "Otherwise it would have been crushed by now. If it's really headlight glass, this could be an important piece of evidence. You're the forensics investigator; can you tell?"

OBJECTIVE

Determine the density of a large glass sample by using Archimedes' principle.

MATERIALS

- suspect glass sample
- tweezers
- water

EQUIPMENT

- balance, mechanical with support platform feature
- beaker, 250 mL or 400 mL
- scissors

SAFETY

- Always wear safety goggles and a lab apron to protect your eyes and clothing.
- Wear protective gloves when handling glass fragments and handle them with caution to avoid cuts and punctures. In the event of a cut or puncture from glass or scissors, notify your teacher immediately.

Procedure

One way to identify headlight glass is to check its density. Typical density values for headlight glass range from 2.47 g/mL to 2.63 g/mL. To figure out the density of the glass sample, you'll need to measure the sample's mass and its volume (using the equation *density = mass/volume*). Because the glass to be tested is rather large and has an unusual shape, its density cannot be determined by measuring the fragment's volume by water displacement. Instead, to determine the density of the glass you should use Archimedes' principle, which states that an object immersed in a fluid will be buoyed upward by a force equivalent to the weight of the fluid the object displaces. In other words, the object will appear to weigh less in the fluid by an amount equal to the weight of the liquid volume the object displaces.

Note that the buoyant force does not depend on the properties of the submerged object itself, but only on the weight of the fluid displaced. Cubes of aluminum, glass, lead, or any other material of the same volume (but with very different weights) will be buoyed upward with the same force.

To determine the density of the glass sample using Archimedes' principle, you will need to know that weight is proportional to mass. You can easily determine the mass of the glass with a balance. According to Archimedes' principle, the apparent loss of weight when the glass is measured in water will be the weight of the water displaced by the glass piece. The volume of this water will be the same as the volume of the glass piece. Because water's density is 1 g/mL, 1 g of water takes up a volume of 1 mL.

The density of the glass piece can be determined using the equation below.

$$\text{density of glass sample} = \frac{\text{mass of glass sample in air}}{\text{mass of glass in air} - \text{apparent mass of glass in water}}$$

1. Create a procedure to test the density of the glass fragment to determine if the glass is part of the crime scene, using Archimedes' principle and the lab materials provided. See the introduction pages on Density, pp. 17–22, for hints. Show your procedure to your teacher for approval.

2. If your procedure is approved, carry out the experiment you have designed. Create a data table that clearly displays your calculations and results.

Postlab Questions

1. What is the density of the glass sample?

   Answers will vary based on the unknown glass samples you provide for the

   students.

2. What statement can be made based on the density of the glass sample?

If the density of the glass sample matches the density for headlight glass or

for safety glass, then the glass sample may be from the headlight of the vehicle

that hit Shanna's car. If the density value does not match the density value for

headlight glass, the glass sample is most likely not related to this accident.

3. What factors could affect the accuracy of your results? How could your procedure have been improved in order to control for these factors?

Student answers may vary. Possible answers include: inaccuracies in the

balance could have resulted in an inaccurate density determination; weighing

several times with different balances and averaging the results would

improve accuracy in this case.

4. Later, at a trial, a defense attorney is cross-examining you: "But does your experiment prove whether this glass came from the car?" How would you defend your results and the conclusions you drew from them?

The experiment alone may disprove whether the glass came from the head-

light of a certain car, but other sources of evidence would be necessary to

prove the glass did come from a certain car.

5. What kinds of further tests might be conducted on the glass to gather more information about the source of the glass sample?

If the suspect glass is shown to be headlight glass and a suspect vehicle is

located, one could determine if the headlight glass from the suspect vehicle

and the glass from the experiment have the same densities. Trying to match

the glass sample by fitting it to the suspect's damaged headlights is another

possible test. One could try to match the tire tracks of the suspect's vehicle

to the tire tracks at the scene of the accident. One could also compare the

paint left on Shanna's car by the car that hit her to the paint of the suspect's

vehicle. Still another test might be to compare the points of impact between

Shanna's car and the suspect's car.

The Parking Lot Collision (density determination using the method of suspension)

Teacher Notes

TIME REQUIRED   one 50-minute class period

LAB RATINGS   Easy ← $\overset{1}{\quad}\overset{2}{\quad}\overset{3}{\quad}\overset{4}{\quad}$ → Hard

    Teacher Preparation–2
    Student Setup–2
    Concept Level–3
    Cleanup–2

SKILLS ACQUIRED

    Designing experiments
    Experimenting
    Collecting data
    Inferring
    Interpreting
    Measuring
    Organizing and analyzing data
    Communicating

SCIENTIFIC METHODS

Make Observations Students will make observations and collect data in order to answer the experimental question.

Analyze the Results Students will analyze their data in a systematic fashion.

Draw Conclusions Students will draw conclusions from their data in order to determine the answer to the experimental question.

Communicate the Results Students will clearly communicate their conclusions based on the outcome of the experiment.

MATERIALS (PER LAB GROUP)

• bottle, amber, for liquid disposal and recovery

• glass chip, side lengths should be about 3 mm

• test tube, 5 mL, with size 00 stopper

• water, distilled and deionized, in dropper bottle

• ZnI_2 solution, saturated, in dropper bottle (see below)

Zinc iodide is a corrosive material which presents a dust hazard. Because inhalation should be avoided, preparation in the hood is recommended. Use appropriate gloves, aprons, and goggles when preparing this chemical.

To prepare 100 mL of ZnI_2 stock solution, dissolve 221 g of ZnI_2 in 50 mL of distilled and deionized water. Then add a small piece of mossy zinc to the solution to prevent oxidation of the iodide to iodine. Before using the solution, check the value of the stock solution density by dividing its mass by its volume. Because the solution will be slightly yellow, the analysis is best performed in direct sunlight so that the glass chip may be seen more easily. Upon completion of the lab, place all the liquids into a large beaker. Simply allow the water to evaporate from the beaker in a hood until the stock solution is again saturated. Check the density of the stock solution before reusing. The ZnI_2 solution is light sensitive, so store it in an amber bottle.

SAFETY CAUTIONS

Because glass fragments are manipulated, there is the potential for students to be cut or punctured by glass. Students should wear appropriate protective gloves and work with caution when handling glass fragments. As usual, appropriate protective goggles and aprons should be worn in the event an accident may produce flying projectiles.

Warn students to use appropriate protective gear in order not to get any ZnI_2 solution on their skin or in their eyes because it is corrosive. Accidents should immediately be reported to the instructor. Caution students to avoid inhaling vapors and use the solution only in well ventilated areas or in the hood.

DISPOSAL

Notify janitorial staff when disposing of broken glass. Carefully consult the supplier's MSDS (Material Safety Data Sheet) for specific warnings, emergency procedures, handling instructions, and directions for proper disposal of ZnI_2. Although the initial cost of ZnI_2 might be of concern, each lab group will use at most a few milliliters of saturated stock solution, and it is possible to reuse the chemical simply by allowing the water to evaporate and then re-dissolving the recovered solid into a saturated stock solution. This should be done in a hood using appropriate chemical safety techniques. If disposal becomes necessary at a later time, follow MSDS disposal procedures.

NOTES ON TECHNIQUE

The usual method of determining density by dividing the mass of an object by its volume is not possible in this instance because of the very small mass and also the irregular shape of the glass chip to be tested. The method of suspension, sometimes termed the method of flotation, is a practical alternative. The density of the chip matches the density of the solution at the point where a glass chip neither sinks nor floats in the solution in which it is immersed. It is then possible to find the density of the solution using weighted averages as shown in the sample calculation on page 27d.

While this technique of determining density is the one most frequently used in crime labs, it can also be the most expensive. Bromoform ($d = 2.89$ g/mL) and bromobenzene ($d = 1.49$ g/mL) are both clear liquids and are mutually miscible. These properties make them the preferred choice for forensic density analysis, but they are hazardous, costly, and difficult to dispose of. The materials used in this activity are a much safer alternative.

Although the glass will not float on the liquid's surface, it can easily be seen to migrate to the top of the liquid. If headlight glass is used as the suspect sample, the glass density (approximately 2.60 g/mL) will be less than the density of the ZnI_2 solution (2.73 g/mL). Therefore, the glass will move upward in the solution. At that point, add only one drop of distilled water to the test tube, place the cap on the test tube, thoroughly mix the liquids, and again observe the movement of the glass chip in the mixture. The goal is to have the glass chip neither float on the top of the liquid nor sink to the bottom, but rather remain suspended in the middle of the liquid.

When the chip is suspended, the density values of the chip and liquid mixture are equal. If the glass moves toward the solution's surface, it is less dense than the solution, in which case the liquid must be made less dense by adding a drop of distilled water. If the chip sinks, it is more dense, and a drop of ZnI_2 solution should be added to increase the liquid's density. It is imperative that the liquids are thoroughly mixed after the addition of each drop and that the exact number of drops to achieve suspension are counted.

TIPS AND TRICKS

Because most types of glass are closer in density to that of the ZnI_2 solution, the analysis should begin by the student placing the glass chip into the test tube and then adding enough ZnI_2 drop by drop until the glass chip can clearly be seen moving to the top of the liquid level.

Students will most likely require assistance to derive the weighted-average equation for determining the density. It is easy to change the outcome of the case by varying the type of glass chips given to each lab group or class. Window-pane, headlight, ophthalmic lens, and other types of glass are easily obtained, stored, and reused. Because the individual glass chips used will vary in size, each lab group's results will differ. The drop size obtained from regular droppers is relatively consistent, but the calibration and use of a pipette would increase accuracy. Students should be instructed that if one drop of water causes the chip to move one way, and a drop of the other liquid causes movement in the opposite direction, they are to disregard the very last drop and count the previous one as a half drop. Density values to within ±0.04 g/mL of the accepted values are readily achieved with this method.

SAMPLE DATA AND CALCULATION

For a small chip of headlight glass, around 25 drops of ZnI_2 solution and 2 drops of distilled water will be required to achieve suspension (it is recommended that you try out the suspension beforehand to verify).

d = density

x = drops ZnI_2 (d = 2.73 g/mL)

y = drops distilled water (d = 0.998 g/mL)

$$d = \frac{(x \times 2.73) + (y \times 0.998)}{x + y}$$

$$d = \frac{(25 \times 2.73) + (2 \times 0.998)}{27} = 2.60 \text{ g/mL}$$

Forensics Lab) INQUIRY LAB

The Parking Lot Collision

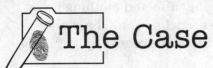 The Case

Mr. Schmitt, the school security guard, got there just in time to separate Nikki and Amanda, who were still screaming at each other.

"I saw you, Amanda. Don't deny it!"

"What?! I didn't do anything, Nikki!"

Nikki turned to Mr. Schmitt. "As I was coming out the doors of the C Building, I saw her back her pickup truck into my car. She even got out, looked at what she had done, and then drove off!"

Mr. Schmitt turned to look at Nikki's car. The hood had buckled, the grille was pushed in, and both headlights were broken.

"Yeah, it's too bad your little car got hurt," said Amanda. "But I didn't do it. Come here and I'll prove it."

They walked over to Amanda's truck, three rows over.

"There, not a scratch on it," said Amanda. "You're just mad at me because I made the softball team and you didn't!"

"Oh yeah?! Well if you…"

"Quiet!" Yelled Mr. Schmitt. "I need to concentrate."

Surprised, the two girls obeyed. He then ran his finger gently along the bumper of Amanda's truck and stopped. "Aha!" he said, holding up a bit of glass. "This may tell us what we need to know. From now on, your case will be handled by the Student Dispute Tribunal, and their Forensic Analysts."

OBJECTIVE

Determine the density of a small glass chip by the method of suspension.

MATERIALS

- dropper bottles of distilled water and saturated ZnI_2 solution
- glass sample

EQUIPMENT

- test tube, 5 mL, with size 00 stopper
- amber bottle for liquid disposal and recovery

SAFETY

- Always wear safety goggles and a lab apron to protect your eyes and clothing.

- Wear protective gloves when handling glass fragments and handle them with caution to avoid cuts and punctures. In the event of a cut or puncture from glass samples, notify the instructor immediately

Name _____ Class _____ Date _____

The Parking Lot Collision *continued*

- In the event a chemical gets on skin or clothing, wash the affected area immediately at the sink with copious amounts of water, keeping affected clothing away from skin. In the event of a chemical spill, notify your teacher immediately. Spills should be cleaned up promptly as directed by the instructor.

Procedure

Your task as Forensic Analyst is to determine the density of the piece of glass retrieved from the bumper of the suspect vehicle. Because the piece is too small to allow direct measurements of mass and volume, you will use the *method of suspension*, or *flotation*.

When an object is placed in a solution and the object neither sinks to the bottom nor floats to the top, but instead remains suspended, then the density of the object is nearly the same as the density of the solution. If you know the density of the solution at that point, you will therefore know the density of the object.

You will use two clear liquids of different density. Glass densities are closer to the density of the saturated zinc iodide solution, so you will want to start by counting 25 drops of the zinc iodide solution into the test tube with the glass sample in it. Then add water dropwise, keeping track of the number of drops until suspension is achieved. Allow the glass fragments to resettle after each drop. Refer to **Figure 1** on the next page as a guide to achieving suspension of the glass fragment in the mixture. When the glass fragment is suspended, you will be able to determine the density of the mixture by calculating a *weighted average* from the densities of the two liquids (based on the number of drops added). If your technique and calculations are accurate, this calculated density of the mixture when the glass piece is suspended will be the same as the density of the glass piece. Density values you will need are listed in **Table 1** below.

TABLE 1 DENSITY VALUES OF TESTED SUBSTANCES

Substance	Characteristic density range
Water at 20°C	0.998 g/mL
Windowpane glass	2.47–2.56 g/mL
Headlight glass	2.47–2.63 g/mL
Ophthalmic glass	2.65–2.81 g/mL
Saturated ZnI_2 solution	2.73 g/mL

1. Create a procedure to test the density of the glass fragment to determine if the glass sample is headlight glass, using the method of flotation and with the lab materials provided. See the introduction pages on Density, pp. 17–22, for hints. Show your procedure to your teacher for approval.

2. If your procedure is approved, carry out the experiment you have designed. Create a data table that clearly displays your calculations and results.

The Parking Lot Collision *continued*

FIGURE 1 DENSITY DETERMINATION BY SUSPENSION

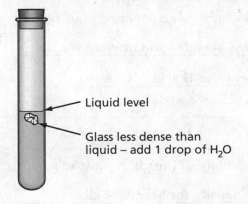

Liquid level

Glass less dense than
liquid – add 1 drop of H_2O

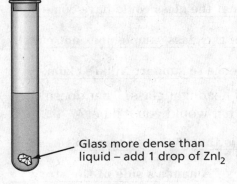

Glass more dense than
liquid – add 1 drop of ZnI_2

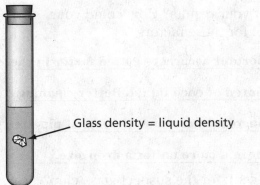

Glass density = liquid density

Postlab Questions

1. What is the density of the glass sample? Show how you calculated your result.

<u>**Answers will vary based on the unknown glass samples you provide for the**</u>

<u>**students. See Sample Data and Calculations in the Teacher Notes.**</u>

_____

_____

2. What statement can you make based on the density of the glass sample?

<u>**If the density of the glass sample matches the density for headlight glass,**</u>

<u>**Nikki's allegation is supported. It is possible that the glass could have come**</u>

<u>**from another source, however. If the density of the glass sample does not match**</u>

<u>**the density of headlight glass, there is no evidence to support Nikki's claim.**</u>

4. Amanda disputes the testing: "So what if it *is* headlight glass? That doesn't prove I did it! There's no point to the test." What would you tell her?

<u>**The main reason why the test is worthwhile is that if the glass wasn't head-**</u>

<u>**light glass, that would give evidence in favor of Amanda's side of the story.**</u>

5. What factors could affect the accuracy of your results? How could your procedure have been improved to control for these factors?

<u>**Student answers may vary. The most important accuracy-related factor in the**</u>

<u>**experiment was the number of drops counted of each liquid. Better cleaning**</u>

<u>**of the glassware and sample would also increase accuracy. The use of pipettes**</u>

<u>**rather than eyedroppers might also result in a more uniform drop size.**</u>

6. How would you interpret a value of 2.50 g/mL for the suspect glass chip?

<u>**Glass of density 2.50 g/mL means the glass could have been a part of a win-**</u>

<u>**dowpane or a headlight. Other tests would be necessary because the density**</u>

<u>**of the glass sample alone cannot absolutely prove Nikki's claim.**</u>

7. What other kinds of tests might be conducted on the glass to help determine the outcome of the case?

<u>**If the suspect glass is shown to be headlight glass, a test could be run on a**</u>

<u>**glass sample from a headlight on Nikki's car to see if the densities are an**</u>

<u>**exact match.**</u>

The Sports Shop Theft (density determination using a gradient column)

Teacher Notes

TIME REQUIRED one 50-minute class period

LAB RATINGS Easy ◄——1——2——3——4——► Hard

    Teacher Preparation–2
    Student Setup–3
    Concept Level–3
    Cleanup–2

SKILLS ACQUIRED

    Designing experiments
    Experimenting
    Collecting data
    Inferring
    Interpreting
    Organizing and analyzing data
    Communicating

SCIENTIFIC METHODS

Make Observations Students will make observations and collect data in order to answer the experimental question.

Analyze the Results Students will analyze their data in a systematic fashion.

Draw Conclusions Students will draw conclusions from their data in order to determine the answer to the experimental question.

Communicate the Results Students will clearly communicate their conclusions based on the outcome of the experiment.

MATERIALS (PER LAB GROUP)

- water, distilled and de-ionized, ($d = 0.998$ g/mL), in a dropper bottle
- ZnI_2 solution, saturated ($d = 2.73$ g/mL), in a dropper bottle
- stoppers, cork, size 0, to stopper the ends of the glass tube (2)
- marker, to indicate layers
- glass samples, crime scene and suspect (each about 3 or 4 mm on each side)
- ¼″ inside diameter glass tubing, about 25 cm in length
- ring stand with utility clamp
- graduated cylinder, 10 mL, for mixing the liquids
- pipettes
- amber bottle for liquid disposal and recovery

Zinc iodide is a corrosive material which presents a dust hazard. Because inhalation should be avoided, preparation in the hood is recommended. Use appropriate gloves, aprons, and goggles when preparing this chemical.

To prepare 100 mL of ZnI_2 stock solution, dissolve 221 g of ZnI_2 in 50 mL of distilled and deionized water. Then add a small piece of mossy zinc to the solution to prevent oxidation of the iodide to iodine. Before using the solution, check the value of the stock solution density by dividing its mass by its volume. Because the solution will be slightly yellow, the analysis is best performed in direct sunlight so that the glass chip may be seen more easily. Upon completion of the lab, place all the liquids into a large beaker. Simply allow the water to evaporate from the beaker in a hood until the stock solution is again saturated. Check the density of the stock solution before reusing. The ZnI_2 solution is light sensitive, so store it in an amber bottle.

SAFETY CAUTIONS

Because glass fragments are manipulated, there is the potential for cuts or punctures. Students should wear appropriate protective gloves and work with caution when handling glass fragments. As usual, appropriate protective goggles and aprons should be worn in the event an accident may produce flying projectiles.

Warn students to use appropriate protective gear in order not to get any ZnI_2 solution on their skin or in their eyes because it is corrosive. Accidents should immediately be reported to the instructor. Caution students to avoid inhaling vapors and use the solution only in well ventilated areas or in the hood.

DISPOSAL

Notify janitorial staff when disposing of broken glass. Carefully consult the supplier's MSDS (Material Safety Data Sheet) for specific warnings, emergency procedures, handling instructions, and directions for proper disposal. Although the initial cost of ZnI_2 might be of concern, each lab group will use at most a few milliliters of saturated stock solution, and it is possible to reuse the chemical simply by allowing the water to evaporate and then re-dissolving the recovered solid into a saturated stock solution. This should be done in a hood using appropriate chemical safety techniques. If disposal becomes necessary at a later time, follow MSDS disposal procedures.

NOTES ON TECHNIQUE

The usual method of determining density by dividing the mass of an object by its volume is not possible in this instance because of the very small mass and also the irregular shape of each glass chip to be tested. The method of suspension, sometimes termed the method of flotation, is a practical alternative. The density of the chip matches the density of the solution at the point where a glass chip neither sinks not floats in the solution in which it is immersed. This lab makes use of a density gradient column, in which 1.5 mL solutions of different densities are made and, starting with the most dense solution, poured separately into the column. A density gradient column makes it simple to compare the densities of two irregularly shaped objects: if they are of the same density, they will settle at the same layer; no further calculation is needed.

Because the solution will be slightly yellow, the analysis is best performed in direct sunlight so that the glass chips may be seen more easily. Instruct students not to squirt liquid into the tube. Instead, they should place the end of the dropper along the inner wall of the tube, and slowly release the liquid into the tube. Each layer should be added as gently as possible to prevent mixing between the layers. The glass tube should be supported with the ring stand and clamp, and handled as little as possible in order to avoid mixing of the density layers.

Upon completion of the lab, place all the liquids into a large beaker, recover the glass chips, and simply allow the water to evaporate in a hood until the stock solution is again saturated. Check the density of the stock solution before reusing and, because it is light sensitive, store in an amber bottle.

TIPS AND TRICKS

Students will most likely require assistance in deriving an equation for determining the density of each layer in the column. It is easy to change the outcome of the Case by varying the type of suspect and crime scene glass chips given to each lab group or class. Window-pane, headlight, ophthalmic lens, and other types of glass are easily obtained, stored, and reused. The 25 cm lengths of glass tubing can accommodate five different 1.5 mL layers, so the 1.5 mL layer volume should be suggested to the students. Layer densities and liquid volumes would be as follows:

Density (g/mL)	mL ZnI_2	mL H_2O
2.50	1.30	0.20
2.55	1.34	0.16
2.60	1.39	0.11
2.65	1.43	0.07
2.70	1.47	0.03

The drop size obtained from regular droppers is relatively consistent, but calibration and use of a pipette would increase accuracy in the preparation of the layers in the column. The use of a calibrated pipette can greatly increase accuracy because of the small volumes of water required to make layers of specific density. If using burettes instead of glass tubing, larger layer volumes will be required. Simply use multiples of the "recipe" for the 1.5 mL layers in the table above. If a greater number of layers is desired, amounts of each liquid required may be determined by substituting desired d values into the equation for x below.

An interesting calibration of the layers may be accomplished by simply dropping crystals of known density into the column. They should migrate to the layer matching their density. Materials appropriate for this approach are as follows (many of these chemicals are hazardous, so MSDS (Materials Safety Data Sheets) should be consulted before use):

$NaClO_3$ $\quad d = 2.49$ g/mL

$KClO_4$ $\quad d = 2.52$ g/mL

NaF $\quad d = 2.56$ g/mL

$MgSO_4$ $\quad d = 2.66$ g/mL

$KMnO_4$ $\quad d = 2.70$ g/mL

$AgClO_4$ $\quad d = 2.80$ g/mL

SAMPLE DATA AND CALCULATION

To compute the density for a layer, the following equation is used.

d = layer density

x = volume, in mL, of saturated ZnI_2 solution ($d = 2.73$ g/mL)

$1.50 - x$ = volume, in mL, of distilled water ($d = 0.998$ g/mL)

$$d = \frac{2.73x + (0.998)(1.50 - x)}{1.50}$$

Consequently, the expression for x, the volume of ZnI_2 to be used in a layer, becomes

$$x = \frac{1.50d - 1.497}{1.732}$$

Using this equation, students may determine the densities that they want for each layer, substitute each desired d value, and compute x (volume of ZnI_2 solution) and $1.50 - x$ (volume of water) for each layer.

The Sports Shop Theft

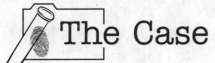 ## The Case

"Thank heavens you're here!" Mr. Antony said as Officer Lewis drove up to the upscale sporting-goods store. Mr. Antony pointed to the back door, which was slightly ajar and had one pane of glass broken.

As Officer Lewis walked up, she bent over and scooped a few bits of broken glass from the door into an evidence envelope. "They broke the glass so they could get the door unlocked," she told Mr. Antony. "Is any of your merchandise missing?"

Mr. Antony ran his eyes across the store's displays. "I don't think so," he said, puzzled. Then he glanced from one wall to the other. "Oh, no," Mr. Antony moaned. "We had framed autographed jerseys and collector edition posters of a lot of great ball players, but they're gone! We'd even screwed the displays into the wall," he added. As she got closer, he spotted the screws scattered on the floor underneath the holes.

"Mr. Antony, whoever did this planned ahead and seems to have known the store. Is there anyone who might consider you an enemy?"

"Well, last week I had to fire a young man named Shawn. He'd been selling all his friends cross-trainers and basketball shoes for half-price."

Later, Officer Lewis visited Shawn, advising him of his rights. "You can wait for me to come back with a search warrant," Lewis said, "but I need to see your shoes."

"I've got nothing to hide," said Shawn. "None of these shoes are the kind that Antony sells in his little 'boutique.'" He handed the officer three pairs of shoes.

Now, Officer Lewis is standing in front of you with two evidence envelopes. "This envelope contains glass from the crime scene. That envelope has glass from the bottom of a suspect's shoes. If you can show they're the same, we may be able to solve this case."

OBJECTIVE

Determine the comparative densities of two small glass chips by the method of suspension in a gradient column.

MATERIALS

- dropper bottles of distilled water (d = 0.998 g/mL) and saturated ZnI_2 solution (d = 2.73 g/ml)
- size 0 cork stoppers
- markers
- crime scene and suspect glass samples

EQUIPMENT

- glass tubing, about 25 cm in length, ¼″ inside diameter
- graduated cylinder, 10 mL
- pipette
- amber bottle, for liquid disposal and recovery
- ring stand with utility clamp

SAFETY

- Always wear safety goggles and a lab apron to protect your eyes and clothing.
- Wear protective gloves when handling glass fragments and handle them with caution to avoid cutting or puncturing yourself. If you cut or puncture yourself from glass samples, notify your instructor immediately.
- In the event a chemical gets on skin or clothing, wash the affected area immediately at the sink with copious amounts of water, keeping affected clothing away from skin. In the event of a chemical spill, notify your teacher immediately. Spills should be cleaned up promptly as directed by the instructor.

Procedure

You will use the density gradient column method for this analysis. This method uses glass tubing as a column, which will contain several layers of liquid, each of a different density. The most dense layer will be found at the bottom and the least dense layer at the column's top. When an object is placed into a liquid and the substance neither sinks nor floats, but instead remains suspended, the densities of the object and the liquid are the same. When dropped into the column, the suspect and crime scene glass samples will be suspended in the layers of the column that match their densities. If the two samples stop in the same density layer, that would indicate that they have the same density.

Your teacher will discuss with you the nature of the liquids to be used and how many layers your column should contain. See **Figure 1** on the next page for a diagram of the suspension column setup. Differing concentrations of ZnI_2 are commonly used, and after you prepare each layer in a graduated cylinder, pour the layer carefully down the side of the column so that the different layers are not mixed. Mark the top of each layer with a marker. If the suspect glass fragment in this case is determined to be window-pane glass, investigation into the incident will continue. Density values you will need are listed in **Table 1** on the next page.

FIGURE 1 SUSPENSION COLUMN SETUP

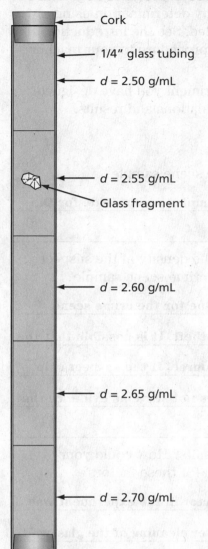

Cork

1/4" glass tubing

d = 2.50 g/mL

d = 2.55 g/mL

Glass fragment

Glass chip location shown
indicates d = 2.55 g/mL.

d = 2.60 g/mL

d = 2.65 g/mL

d = 2.70 g/mL

TABLE 1 DENSITY VALUES OF TESTED SUBSTANCES

Substance	Characteristic density range
Water at 20°C	0.998 g/mL
Window-pane glass	2.47–2.56 g/mL
Headlight glass	2.47–2.63 g/mL
Ophthalmic glass	2.65–2.81 g/mL
Saturated ZnI_2 solution	2.73 g/mL

1. Create a procedure to test the comparative densities of the glass fragments in order to solve the case, using the method of density determination using a gradient column and with the lab materials provided. See the introduction pages on Density, pp. 17–22, for hints. Show your procedure to your teacher for approval.

2. If your procedure is approved, carry out the experiment you have designed. Create a data table that clearly displays your calculations and results.

Postlab Questions

1. What is the density of the glass sample? the crime scene sample?

   Answers will vary based on the unknown glass samples you provide for the

   students.

2. What can you conclude about the case based on the density of the suspect glass sample as compared with the density of the crime scene sample?

   If the density of the suspect chip matched the value for the crime scene

   glass, then Shawn may have been involved in the theft. It is possible that the

   glass in his shoe could have come from another source. If the suspect chip

   does not match the crime scene glass, then it does not serve as evidence that

   Shawn was present at the crime scene.

3. What factors could affect the accuracy of your results? How could your procedure have been improved in order to control for these factors?

   Student answers may vary. The most important factor in the experiment was

   the accuracy in the number of drops counted. Better cleaning of the glassware

   and sample would also increase accuracy. The use of a pipette rather than

   eyedroppers might also result in a more uniform drop size and better accuracy.

4. How would you interpret a value of 2.50 g/mL for the suspect glass chip?

   Glass of density 2.50 g/mL could be either from a window-pane or headlight.

   Other types of tests would then be necessary to determine the origin of the

   suspect glass (for example, Shawn could have stepped on a headlight glass

   fragment in a parking lot).

Procedure Introduction

A Lesson on Chromatography

Green paint has been thrown all over several walls of a local business. The police have several suspects and have found some green paint at each of the suspects' homes. How can the police determine if any of the suspect samples match the paint thrown on the walls?

BACKGROUND

Chromatography is a technique used to separate a mixture of different substances based on the polarity of the molecules of the substances. It was originally developed in 1903 by a Russian botanist, Mikhail Tswett, who used this approach to separate colored plant pigments. The word means "to write with colors" and comes from the Greek words *chroma*, "color," and *graphein*, "to write."

There are several types of chromatography, each one depending on the nature of the substance in a mixture. In most types of chromatography, *polarity* is the basic principle at work. Water is the best example of a polar substance: its molecules have an uneven distribution of electrical charge, which means that one side of a water molecule strongly attracts the other side of the water molecule next to it. Polar substances have strong attraction to each other because of their molecules' ability to attract one another in this way, whereas nonpolar substances, such as oils, have a much stronger attraction for each other than for polar substances.

HOW CHROMATOGRAPHY WORKS

All kinds of chromatography involve two phases in contact with each other: a *mobile phase* and a *stationary phase*. **Table 1,** below, shows some examples of what forms these can take in different types of chromatography.

TABLE 1 SOME TYPES OF CHROMATOGRAPHY

Category	Mobile phase	Stationary phase	Separating principle
Paper chromatography	some solvent; can be water, methanol, etc.	paper	polarity of mixture components relative to mobile phase
Thin-layer chromatography	some solvent; can be water, methanol, etc.	silica gel plate	polarity of mixture components relative to mobile phase
Column chromatography	some solvent; can be water, methanol, etc.	powdered adsorbent packed in a glass column	polarity of mixture components relative to mobile phase

A Lesson on Chromatography *continued*

The mobile phase is a solvent that dissolves some or all of the substances to be separated in the mixture. The stationary phase on which the mixture is placed is often some type of solid material such as chromatography paper or a thin layer of a gel coating a glass or plastic plate. As the mobile phase moves through or across the stationary phase, the stationary phase separates the components of the mixture. The result of the process is called a *chromatogram*, named for the colored bands produced when the separation involves a mixture of colored substances.

PAPER CHROMATOGRAPHY

A paper chromatogram is produced by placing a sample of the mixture to be separated near one end of a piece of chromatography paper (the "stationary phase"). The paper is placed into a container so that the paper hangs free without touching the walls of the container. A solvent, the "mobile phase," is placed into the container. The solvent level should not be higher than the sample spot or the mixture will dissolve in the solvent instead of traveling up the paper. A diagram showing the setup for paper chromatography is shown below in **Figure 1.**

FIGURE 1 EXPERIMENTAL SETUP FOR PAPER CHROMATOGRAPHY

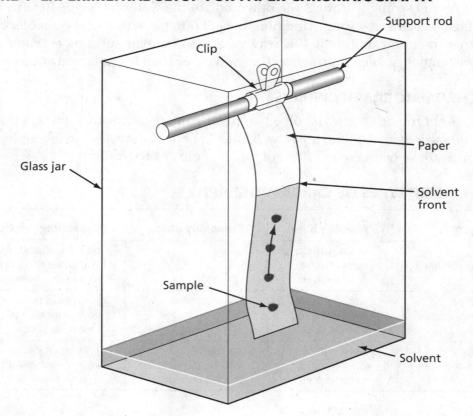

While the paper is in contact with the solvent, the solvent moves upward through the paper. Substances in the sample mixture are dissolved by the solvent and move along the paper with the solvent. As the solvent moves through the paper, the molecules of the dissolved substances are attracted to both the solvent and the paper. The strengths of these attractions are determined by how polar the substance is. Polar substances will have a greater attraction to one of the phases than substances that are nonpolar. Because these attractions are different for each substance in a mixture, each substance moves through the paper at a different rate. The substances with a greater attraction to the solvent move faster and farther along the paper. Those with a greater attraction to the paper move more slowly, and consequently, not as far as other substances. Polar solvents tend to work best with substances that are also polar, such as water-based inks and dyes. A less-polar solvent will tend to work best with nonpolar or weakly polar substances, such as oil-based inks and dyes.

In addition, large molecules tend to move more slowly than small molecules because the large molecules cannot easily pass through the stationary phase material. The result of the differences in attraction and size cause the various substances to separate into distinct areas on the paper. The paper is removed from the solvent before the solvent reaches the end of the paper opposite the solvent. The paper is then allowed to dry, producing a permanent record that can be analyzed or processed further.

For a given mixture, the choice of the solvent used for the mobile phase can affect the results obtained. The solvent used for the mobile phase should dissolve most or all of the components of the mixture. A polar solvent is more likely to give good results with substances that are polar. Likewise, a nonpolar solvent is more likely to give good results with nonpolar substances. Water, methanol, and acetic acid are examples of highly polar solvents. Petroleum ether, cyclohexane, and methylene chloride are examples of solvents that are nonpolar.

HOW TO USE CHROMATOGRAPHY

Paper chromatography is often used as a tool for comparing the makeup of two or more mixtures. If the chromatogram is the same for each mixture, then the mixtures are probably the same. If the chromatograms are different, then the mixtures are probably different. Often, different samples to be tested are placed on the same piece of chromatography paper. This way, the conditions that create the chromatogram are the same for all the mixtures.

Not every solvent will give good results. A chromatogram produced using a particular solvent may drive all the components of the mixture up the paper too quickly, and another solvent may not cause the particular components of a mixture to travel up the paper at all. The right solvent for a mixture will give clear and consistent separation between the mixture's components.

A Lesson on Chromatography *continued*

1. In chromatography, what are the functions of the mobile and stationary phases?

   The mobile phase carries the mixture to be separated by chromatography.

   The stationary phase attracts some of the components of the mixture.

2. Red dye can be made from a mixture of yellow dye and magenta (a deep purplish red) dye. How could you determine if a particular red dye is made from a single dye or from a mixture of yellow and magenta dyes?

   Produce a paper chromatogram of the dye to see if the dye separates into

   yellow and magenta. If it separates, then the dye is a mixture of yellow and

   magenta dyes. If the chromatogram produces only a red spot, then the dye is

   a single red dye.

3. Two mixtures are compared using chromatography. Why is it desirable to place samples of both mixtures on the same paper when making the chromatogram?

   Placing both mixtures on the same paper ensures that both mixtures are

   separated under the same set of conditions. Small differences in the solvent

   or paper and the length of time the paper is in the solvent are all factors

   that can affect the results of the separation.

4. A paper chromatogram is made for a particular mixture. The colored bands that result from the chromatography run are so close to each other that they overlap, making the chromatogram difficult to interpret. What change or changes could be made to increase the separation between the colors? Explain.

   By increasing the length of the paper, the different components of the mix-

   ture will have more time to interact with the paper. Using a different solvent

   with a different degree of polarity may also produce a greater separation.

The Counterfeit Drugs (separation of dyes using strip-paper chromatography)

Teacher Notes

TIME REQUIRED one 50-minute class period

LAB RATINGS   Easy ←——1———2———3———4——→ Hard
Teacher Preparation–1
Student Setup–2
Concept Level–2
Cleanup–2

SKILLS ACQUIRED
Experimenting
Inferring
Interpreting
Communicating

SCIENTIFIC METHODS

Make Observations Students will make observations and collect data in order to answer the experimental question.

Analyze the Results Students will analyze their data in a systematic fashion.

Draw Conclusions Students will draw conclusions from their data in order to determine the answer to the experimental question.

Communicate the Results Students will clearly communicate their conclusions based on the outcome of the experiment.

MATERIALS (PER LAB GROUP)

• beakers, small (2)

• candies, chocolate, brown-colored (20 per class)

• eyedroppers

• food coloring, brown

• paper clips, small

• pencils

• scissors

• strip chromatography paper (2 pieces)

• tape

• water, distilled

SAFETY CAUTIONS

No chemicals are used in this activity. Because glassware and scissors are manipulated, there is the potential for students to cut or puncture themselves. As usual, appropriate protective goggles and aprons should be worn in the event an accident may produce flying projectiles. Notify janitorial staff when disposing of broken glass.

DISPOSAL

All liquids can be safely disposed of down the drain.

NOTES ON TECHNIQUE

This activity is a very simple chromatography experiment that will serve as a suitable introduction to the technique. Chromatography is a means of separating the components in a mixture based on each component's interaction with a stationary phase, in this case the paper, and a moving phase, in this case water. Each of the three—dye components, paper, and solvent—contains polar molecules, and the differences in polarity between the dye components will result in their separation as they are carried up the paper by the solvent. Chromatography paper is very inexpensive and may be purchased in either pre-cut strips, sheets, or in rolls. The size of the beakers that are used will determine the chromatogram lengths.

TIPS AND TRICKS

To prepare a dye sample for the "generic drug," place about 20 brown-colored chocolate candies into a 50 mL beaker, and add a teaspoonful of water. The candy consists of an outer dye coating, a white sugar coating, and a chocolate interior. Stir the candy until the brown coating is dissolved in the water, and then pour this liquid into another small beaker. The dye coating of the brown candies contains blue and orange dyes, which should separate out upon chromatographic testing and enable students to identify the "drug" as the "generic" version, as described in the case description.

If you choose to provide some lab groups with a dye that will give a negative result (the "patented drug"), you might want to use a brown food coloring. Test the food coloring first to ensure that it does not give a similar result as the dyes from the brown-colored chocolate candies.

It is important that the dye be spotted repeatedly on the chromatography paper to obtain enough pigment to test. Students should allow each spot to dry before applying another one on top of it. A common student error is to let the runs proceed too long, in which case all of the pigments will be carried to the very top of the paper. So it is important to instruct students to remove the paper strips from the beakers when the solvent migrates to within a few centimeters of the pencil and while the pigments are still spread out on the chromatography paper.

Forensics Lab

The Counterfeit Drugs

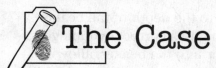 The Case

A pharmaceutical company has been accused of illegally manufacturing and exporting a counterfeit version of a popular blood pressure medication. The Food and Drug Administration (FDA) suspects that the company manufactures a generic version of the patented drug and labels it identically to the real drug. The FDA, along with the Drug Enforcement Administration (DEA), have been monitoring the pricing and sales activity of the company and have now confiscated a sample of the drug being sold by the firm.

There is a difference between the generic and patented versions of the medication: the pigments used in the color coating of the pills. The generic blood pressure pill is a brown color known to contain water-soluble blue and orange pigments, whereas the patented version is coated with a material containing a homogeneous brown dye, also water soluble. The FDA has contracted with your lab to test a sample of brown pigment in water solution by chromatography to see if the pigment is from the generic pills or the patented pills.

OBJECTIVE

Determine by strip paper chromatography if a suspect dye contains a blue pigment.

MATERIALS

- distilled water
- pigment solutions to be tested
- strips of chromatography paper (2)
- tape

EQUIPMENT

- beakers, small (2)
- eyedropper
- paper clips, small
- pencil
- scissors

SAFETY ◈ ◈

- Always wear safety goggles and a lab apron to protect your eyes and clothing.

- Handle all glassware with caution. In the event of a cut or puncture from broken glass or scissors, notify the instructor immediately.

Name _____ Class _____ Date _____

The Counterfeit Drugs *continued*

Procedure

1. Cut a length of strip chromatography paper equal to the depth of the beaker you are using. You may check to make sure the length is correct by taping the top of the paper to a pencil and lowering the paper into the beaker, allowing the pencil to rest across the top of the beaker.

2. Once the paper is cut to the correct length, remove it from the beaker. Clip a paper clip to the bottom of the paper to keep the paper hanging straight down while it is in the beaker (keep the pencil attached to the top of the paper). Using an eyedropper, spot some of the suspect dye onto the chromatography paper about 2 cm above where the water level will be in the bottom of the beaker.

3. Allow the dye to dry for a few minutes, and then repeat this spotting process several times (at the same place on the paper each time) to build up a concentration of pigment that will yield good results.

4. Now place 10 mL of distilled water into the beaker. It is important to have the test spot be 2 cm above the water level, and not immersed, so make sure that the dye to be tested does not come into contact with the water.

5. To conduct a chromatographic run, place the paper length into the beaker with the pencil resting across the top of the beaker, and allow the water to move up the paper by capillary action and separate the pigments in the dye. This process may take several minutes, so plan accordingly.

6. When the water migrates to a few millimeters from the top of the paper, stop the run by removing the pencil and attached paper from the beaker. Empty the water from the beaker, place the pencil back across the top of the beaker, and allow the chromatogram to dry overnight.

7. Dispose of all materials as instructed by your teacher.

Analysis

1. **Explaining Events** Why do the pigments of a dye separate out on the chromatography paper?

   Molecules of pigments in the dye, paper, and water are all polar, so the

   pigments will travel along the paper with the water. The different dyes have

   different attractions to the water and paper, however, so they travel up the

   paper to different heights.

The Counterfeit Drugs *continued*

2. Analyzing Methods Why is it important that tests in cases like this one be carried out such that the analysts do not know what products they are testing?

If the analysts do not know which samples are being tested, the results can

be considered to be free of bias; that is, the outcomes were determined only

by the natures of the samples, not by what the analysts were expecting to

find.

Conclusion

1. Drawing Conclusions Is the sample you tested from the generic drug or the patented drug? How can you tell?

Results will depend on the dye solutions you provide to the lab groups. Dye

solutions that yield the blue and orange components should be considered

from the "generic" version of the drug. Those solutions that did not produce

the blue and orange components (for instance, if you used a food coloring

dye as the "patented" version), should be considered as not proven to have

been illegally manufactured.

Extension

1. Research Research and report on science-related careers in the FDA or DEA.

The Athletic Rivals (ink analysis using paper chromatography)

Teacher Notes

TIME REQUIRED   two 50-minute class periods

LAB RATINGS

Easy ← 1 2 3 4 → Hard

Teacher Preparation–2
Student Setup–3
Concept Level–3
Cleanup–2

SKILLS ACQUIRED

Designing experiments
Experimenting
Inferring
Interpreting
Communicating

SCIENTIFIC METHODS

Make Observations Students will make observations and collect data in order to answer the experimental question.

Analyze the Results Students will analyze their data in a systematic fashion.

Draw Conclusions Students will draw conclusions from their data in order to determine the answer to the experimental question.

Communicate the Results Students will clearly communicate their conclusions based on the outcome of the experiment.

MATERIALS (PER LAB GROUP)

- capillary tubes, open-ended (3)
- chromatography paper (6 strips)
- dropper bottle of distilled water
- dropper bottle of methanol
- graduated cylinder, 10 mL
- paper clips (12)
- rubber stoppers, no-hole, #4 (6)
- single hole puncher
- scissors
- spot plates
- suspect pens (2) (black felt tips work best because most are heterogeneous)
- test tubes, 25×200 mm (6)
- test tube rack
- a threatening note
- thumbtacks (6)

The Athletic Rivals *continued*

SAFETY CAUTIONS

Methanol (methyl alcohol) can be absorbed through the skin during prolonged exposure. Wear protective gloves and goggles when handling this chemical and remember to check students for allergy to latex if latex gloves are to be used. This alcohol is flammable, a vapor hazard, and may cause blindness with ingestion or repeated/prolonged exposure. To avoid inhalation, handle only in a well-ventilated area or hood.

Because glassware and scissors are manipulated, there is the potential for students to cut or puncture themselves. As usual, appropriate protective goggles and aprons should be worn in the event an accident may produce flying projectiles. Notify janitorial staff when disposing of broken glass.

In the event a large quantity of methanol is spilled, students should be evacuated from the room until cleanup is complete.

DISPOSAL

Carefully consult the supplier's MSDS (Material Safety Data Sheet) for specific warnings, emergency procedures, handling instructions, and directions for proper disposal of methanol.

NOTES ON TECHNIQUE

Chromatography is a means of separating the components in a mixture based on each component's interaction with a stationary phase, in this case the paper, and a moving phase, the solvent. Each of the three—dye components, paper, and solvent—contain polar molecules, and the differences in polarity of the ink components will result in their separation as they are carried up the paper by the solvent. Chromatography paper is very inexpensive and may be purchased in pre-cut strips, sheets, or in rolls. The size of the test tubes you have will determine stopper size and chromatogram dimensions. Emphasize to students that their six chromatograms must be of the same dimensions.

TIPS AND TRICKS

It will take one complete class period for the student groups to prepare for the runs and another period to actually conduct them. Black, thicker tip, water-soluble markers and permanent markers contain several pigments. It is not obvious by simple examination whether a note has been written with a water-soluble or a permanent marker, so you can have different notes for classes or lab groups.

The chromatograms obtained by directly marking on the paper with the suspect pens will yield the best results. Students should repeatedly spot ink extracted from the note onto the chromatography paper to obtain good results. Students should allow each spot to dry before applying another one on top of it. It is important to be sure that students remove the paper strips when the solvent migrates to within a few millimeters of the top of the paper.

Forensics Lab

The Athletic Rivals

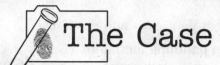 The Case

Jason scowled at the broken windshield glass on the parking lot pavement. "Kyle's gonna pay for this," he muttered. His car had been vandalized sometime during the day.

Rivals for the role of starting quarterback on the football team, Kyle and Jason have been steadily growing more at odds with each other. "Those threatening notes I've gotten—and I know it was Kyle who wrote them—said something like this might happen," he told the campus police officer who had come to the scene, "and now it has. At least we know who did it."

"Hold on a minute," Officer Harlow said, "do you have proof that Kyle wrote the note you're referring to?"

Jason hesitated a moment. "Well, no, but who else could it be?"

"If we can find some way to prove Kyle threatened to do this, then you could file charges against him. Until then, I don't know what else you could do."

"So how do we do that?" Jason asked.

"You give me that threatening note, and I'll talk to Kyle," said officer Harlow.

Officer Harlow made a search of Kyle's locker and found two pens that look like they could have been used to write the note that Jason turned over as evidence. Kyle denies writing the note and any involvement in the vandalism.

The two pens and the note threatening vandalism to Jason's car, have been turned over to you, forensic lab analyst. Your job is to prepare paper chromatograms for Kyle's two suspect pens and for the ink extracted from a threatening note to see if there is a pigment match.

OBJECTIVE

Determine by examination of pigments through chromatography whether a suspect pen could have been used to write a threatening note.

MATERIALS

- open-ended capillary tubes (3)
- chromatography paper (6 strips)
- dropper bottle of distilled water
- dropper bottle of methanol
- segment of threatening note
- suspect pens
- thumbtacks (6)

EQUIPMENT

- 25 × 200 mm test tubes (6)
- test tube rack
- #4 no-hole rubber stoppers (6)
- single hole puncher
- 10 mL graduated cylinder
- spot plate
- scissors
- thumbtacks (6)

SAFETY   ◈

- Always wear safety goggles and a lab apron to protect your eyes and clothing.

- Methanol is poisonous. Wear protective gloves when handling methanol. (Notify your teacher if you are allergic to latex.) Avoid prolonged exposure to vapors and use in the hood or a well-ventilated area, as directed by instructor. Keep methanol away from heat and flames, as it is flammable.

- In the event methanol gets on skin or clothing, wash the affected area immediately at the sink with copious amounts of water, keeping affected clothing away from skin. In the event of a chemical spill, notify the instructor immediately. Spills should be cleaned up promptly as directed by the instructor.

- Handle all glassware with caution. In the event of a cut or puncture from broken glass or scissors, notify the instructor immediately.

Procedure

Some inks are water soluble and others dissolve in methanol, so you will need to make six chromatographic runs: inks from suspect pen 1, suspect pen 2, and the note will each be tested once with water, and once with methanol, as solvents.

1. Cut a length of chromatography paper so that it extends the entire length of a test tube. Bend the top of the paper at a right angle and tack it to the bottom of the rubber stopper. Push a thumbtack through the paper near the bottom to make it hang straight down in the tube so that the solvent can travel upward (take care not to puncture your fingers).

2. Repeat step 1 until you have made six paper/test tube setups. Each paper segment should be of identical length and width, and make sure that the segments do not touch the sides of the test tubes.

3. To extract ink from the note, use a hole puncher to punch about 40 holes in the note from areas containing ink. Place half of these small circles into a spot plate well, and then add a few drops of water (for the water run).

4. Place the rest of the punched holes containing ink into another spot well, and add methanol (for the methanol run) to dissolve and remove the ink from the paper.

5. Using a capillary tube, remove some of the water-based note ink solution. Spot this solution onto one of the paper lengths, about 2 cm above where the solvent level will be. Repeat this spotting process (allowing each spot to dry before applying another on top of it) several times to build up a concentration of note pigment that will yield good results.

6. To prepare chromatograms for the two suspect pens, simply mark spots about 3 mm in diameter directly onto a chromatography strip for each pen, about 2 cm above where the solvent level will be, as in the chromatogram prepared by spotting in step 5.

7. Repeat steps 5 and 6 to spot the methanol-based ink solution and ink from the two suspect pens onto three additional strips of chromatography paper, respectively. Be sure to keep track of which chromatogram was spotted with the water-based ink solution and which was spotted with the methanol-based ink solution.

8. Place 8 mL of water into each of three test tubes. Place 8 mL of methanol into each of another three test tubes.

9. A diagram of the chromatography setup is shown in **Figure 1** below. Place the stoppers with attached paper lengths into the test tubes. Be sure to place the methanol-ink spotted chromatograph into one of the test tubes with methanol as the solvent, and the water-ink spotted chromatograph into one of the test tubes with water as the solvent. Make sure the ink spot on each chromatography strip does not touch the solvent (water or methanol).

10. Allow the solvent to move up the paper and separate the pigments in the ink (this process may take up to 30 minutes). Stop the runs by removing the papers from the test tubes when the solvent level has risen to within a few centimeters of the bottom of the stoppers, and allow the chromatograms to dry overnight. Dispose of materials as directed by your teacher.

FIGURE 1 PAPER CHROMATOGRAPHY SETUP

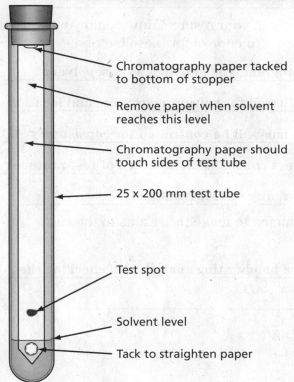

Chromatography paper tacked to bottom of stopper

Remove paper when solvent reaches this level

Chromatography paper should touch sides of test tube

25 x 200 mm test tube

Test spot

Solvent level

Tack to straighten paper

The Athletic Rivals *continued*

Postlab Questions

1. Are the pigments of the note ink and the order of these pigments consistent with the pigments and order of pigments of the suspect pen inks?

Student answers will vary based on samples provided. Note that it will

not be the concentration of pigments that indicates similarity between note

ink and suspect pen ink, but rather the type and order of pigments on the

chromatograms.

2. If the note pigments match one of the suspect pens, how exactly does this help answer the question of whether Kyle wrote the note?

If the note pigments match the pigments of the suspect pens, then investiga-

tors should next determine if the suspect pen is a common brand. If the pen

is a common brand, then Kyle cannot be strongly linked to the notes. If the

pen is of a rare brand, that would be a much stronger indication that Kyle

wrote at least one of the notes.

3. What factors could affect the accuracy of your results? How could your procedure have been improved in order to control for these factors?

Student answers may vary. Chromatogram type and dimensions, solvent

amounts, layers of ink used, time allowed for solvent migration, and location

of solvent front at the end of a run must all be controlled for consistency

throughout testing. Defense attorneys commonly attack lack of test protocol

to create reasonable doubt in the validity of test results.

4. What additional tests could be conducted to make the results of the case more certain?

Additional tests may include doing a handwriting analysis and checking the

threatening note for fingerprints.

Forensics Lab

The Questionable Autograph (ink analysis using thin layer chromatography)

Teacher Notes

TIME REQUIRED two 50-minute class periods

LAB RATINGS Easy ←—1——2——3——4—→ Hard
   Teacher Preparation–2
   Student Setup–3
   Concept Level–3
   Cleanup–1

SKILLS ACQUIRED
   Designing experiments
   Experimenting
   Inferring
   Interpreting
   Communicating

SCIENTIFIC METHODS

Make Observations Students will make observations and collect data in order to answer the experimental question.

Analyze the Results Students will analyze their data in a systematic fashion.

Draw Conclusions Students will draw conclusions from their data in order to determine the answer to the experimental question.

Communicate the Results Students will clearly communicate their conclusions based on the outcome of the experiment.

MATERIALS (PER LAB GROUP)

- bottles, glass, 2 oz., with cap (6)
- capillary tubes, open-ended
- dropper bottle of distilled water
- dropper bottle of methanol
- graduated cylinder, 10 mL
- paper strips with pen markings (with a different pen from the autograph samples if you want the test of differences to turn out positive)
- puncher, single-hole
- rulers for scoring
- thin layer chromatography plates
- teasing needles or pushpins for scoring and scraping gel from the TLC plates
- ticket stub with autograph written in pen

SAFETY CAUTIONS

Methanol (methyl alcohol) can be absorbed through the skin during prolonged exposure. Wear protective gloves and goggles when handling this chemical and remember to check students for allergy to latex if latex gloves are to be used. This alcohol is flammable, is a vapor hazard, and may cause blindness with ingestion or repeated/prolonged exposure. To avoid inhalation, handle only in a well-ventilated area or hood.

Because glassware and teasing needles are manipulated, there is the potential for students to cut or puncture themselves. Handle sharp objects with care. As usual, appropriate protective goggles and aprons should be worn in the event an accident may produce flying projectiles. Notify janitorial staff when disposing of broken glass.

In the event a large quantity of methanol is spilled, students should be evacuated from the room until cleanup is complete.

DISPOSAL

Carefully consult the supplier's MSDS (Material Safety Data Sheet) for specific warnings, emergency procedures, handling instructions, and directions for proper disposal of methanol.

NOTES ON TECHNIQUE

Chromatography is a means of separating the components in a mixture based on each component's interaction with a stationary phase, in this case the solid gel coating on the TLC plates, and a mobile phase, the solvent. Each of the three— dye components, gel, and solvent—contain polar molecules, and the differences in polarity of the ink components will result in their separation as they are carried up the TLC plate by the solvent. Full TLC sheets are very inexpensive and may be cut with scissors to dimensions that will fit into the bottles, usually to about the size of microscope slides. The size of the bottles that you have will dictate the optimal dimensions for the plates. The solid gel coating on the sheets has the consistency of flour, and it will scrape off easily.

TIPS AND TRICKS

It will probably take one complete class period for the student groups to prepare for the runs and another period to actually conduct them. Black, thicker tip, water-soluble markers contain several pigments, as do permanent markers. It is not obvious by simple examination whether an autograph has been written with a water-soluble or a permanent marker, so you can have different autographs for classes or lab groups.

It is important that the ink extracted from the autographs be layered to obtain enough pigment to test. In addition, students should let the first spot dry before applying another layer, exactly on top of the first spot. Make sure the students have the test spots situated well above the solvent in the bottles. A common error is for students to allow the runs to proceed too long, in which case all of the pigments will be carried to the very top of the plate and be smeared together. So it is important to instruct students to remove the plates when the solvent migrates to within a centimeter of the top of the plate.

Name _____ Class _____ Date _____

The Questionable Autograph

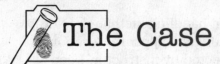 The Case

"I've been ripped off! What can I do now?" Juan complained. "I was so excited to win an online auction for a ticket stub autographed by my favorite ballplayer, but I really don't think it's authentic."

"What makes you think so?" Ms. Pierson asked.

"I heard he signed an endorsement with a certain brand of pen to sign autographs with only their pens, and this doesn't look like it was signed with that type of pen—it looks like it was just signed with a regular ballpoint pen! I contacted the online auction company with my complaint and they told me they were just about to ban the seller because of complaints, and that an investigation of mail fraud was underway. That's when I called you," Juan said.

"You called the right person," Ms. Pierson said. "As a U.S. Postal Inspector, it's my job to investigate all reports of mail fraud. I'll need to take that ticket stub back to the lab so that we can analyze the ink to see if it comes from the right kind of pen. The ticket stub will have to be destroyed in the process, though."

"Well, all right, if that's the only way to get my money back," Juan said, as he reluctantly handed over the evidence.

Ms. Pierson has turned the signed ticket stub over to you, forensic lab analyst, to compare by thin layer chromatography (TLC) the ink from the questioned item with ink from the type of pen the athlete would have used.

OBJECTIVE

Determine by examination of pigments through thin layer chromatography if an autograph was written with a particular type of pen.

MATERIALS

- open-ended capillary tubes (2)
- thin layer chromatography plates (2)
- dropper bottles of distilled water and methanol
- suspect autographs, and ink samples from an authentic pen

EQUIPMENT

- 2-oz. glass bottles with caps (2)
- graduated cylinder, 10 mL
- ruler
- single-hole puncher
- spot plates
- teasing needle or pushpin

SAFETY ◈ ◈ ◈ ◈ ◈

- Methanol is poisonous. Wear protective gloves when handling methanol. (Notify your teacher if you are allergic to latex.) Avoid prolonged exposure to vapors and use in the hood or a well-ventilated area, as directed by instructor. Keep methanol away from heat and flames, as it is flammable.

- In the event methanol gets on skin or clothing, wash the affected area immediately at the sink with copious amounts of water, keeping affected clothing away from skin. In the event of a chemical spill, notify the instructor immediately. Spills should be cleaned up promptly as directed by the instructor.

- Handle all glassware with caution. If you cut or puncture yourself from broken glass or scissors, notify the instructor immediately.

Procedure

Some inks are water soluble and others dissolve in methanol, so you will need to conduct four thin layer chromatography (TLC) runs: two (one for the suspect ink and one for the authentic ink) using water as the solvent, and another two of the same samples using methanol as the solvent. Refer to **Figure 1** below for aid in the TLC setup.

FIGURE 1 THIN LAYER CHROMATOGRAPHY PLATE SETUP

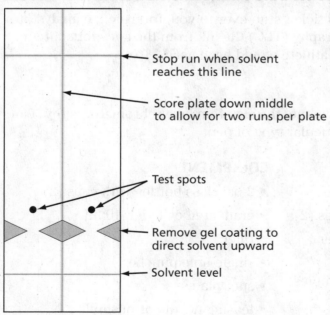

Stop run when solvent reaches this line

Score plate down middle to allow for two runs per plate

Test spots

Remove gel coating to direct solvent upward

Solvent level

The Questionable Autograph *continued*

1. Score two TLC plates in half lengthwise by scratching a line on the TLC gel coating with the teasing needle or pushpin, to allow two runs per plate. Also scratch off some of the plate coating from the sides of the plate below the test spots, as shown in **Figure 1** on the previous page. This will help the solvent to migrate more directly upward so as to achieve better dye separation on the plate.

2. To extract ink from an autograph, use a hole puncher to punch about 40 holes from areas containing ink. Place half of these small circles into a spot plate well and then add a few drops of water to dissolve and remove the ink from the paper.

3. Place the rest of the paper punches containing autograph ink into another spot plate well, and add a few drops of methanol to dissolve and remove the ink from the paper.

4. Repeat steps 2 and 3 with ink samples from the authentic pen.

5. You will use one of the TLC plates to run the suspect and authentic inks side by side with water as the solvent, and the other TLC plate to run the suspect and authentic inks side by side with methanol as the solvent. Use a capillary tube to spot some of the suspect ink onto the TLC plates (about 1 cm above where the solvent level will be in the bottom of bottle. Allow the initial spots to dry, then place more ink on top of the previous spots. Repeat this process a few times to build up a concentration of pigment that will yield good results.

6. Use a different capillary tube to place authentic ink on the other side of the TLC plates in the same way as you did the suspect ink in step 5. Be sure to keep track of which side on each plate is being used for which ink sample.

7. Place 5 mL of solvent into each of the two bottles: one with water and one with methanol. It is important for both runs to have the test spots be 1 cm above the solvent level and not immersed or even touching the solvent initially.

8. Place the TLC plates, as upright as possible, into the bottles, screw on the caps, and allow the solvents to move up the plates and separate the pigments in the ink. Observe the process carefully, as it may take just a few minutes, or up to 30 minutes, for the solvent to move to the top of the plate. When the solvent level has risen to within a centimeter of the top of the plate, remove the plate from the bottle. Allow the TLC plates to dry overnight before examining them.

9. Dispose of materials as instructed by your teacher.

Postlab Questions

1. Are the pigments contained in the suspect and authentic inks similar?

Student answers may vary. Note that it will not be the concentration of

pigments that indicates similarity between authentic ink and suspect ink, but

rather the type and order of pigments on the chromatograms.

2. If the two inks tested are identical, does this prove with certainty that the autograph is authentic?

If the pigments match, investigators should determine if the authentic pen is

a common brand. If it is, then more tests would be made to establish that the

autograph is authentic. If the pen is of a rare brand, and the autograph from

the ticket stub and the authentic ink match, it is more likely that the ques-

tionable autograph is authentic.

3. What factors could affect the accuracy of your results? How could your procedure have been improved in order to control for these factors?

Student answers may vary. Chromatogram type and dimensions, solvent

amounts, layers of ink used, time allowed for solvent migration, and location

of solvent front at the end of a run must all be controlled for consistency

throughout testing. Defense attorneys commonly attack lack of test protocol

to create reasonable doubt in the validity of test results.

4. What additional tests could be conducted in order to make the results of the case more certain?

Additional tests may include handwriting analysis and checking the ticket

stub for fingerprints of the athlete.

Procedure Introduction

A Lesson on Spectroscopy

You enter your chemistry laboratory and find a note left by your research assistant that reads, "Use these for your analysis." There are two flasks containing yellow solutions nearby. One is labeled with the concentration but does not state the identity of the chemical compound, and the other is labeled with the identity of the chemical compound but not the concentration. How would you determine the identity of the first solution and the concentration of the second solution?

BACKGROUND

The technique known as *spectroscopy* was perfected in the 1850s and was publicly announced to the scientific community in 1859. The fundamental principle behind spectroscopy is that each element will emit or absorb a different pattern of electromagnetic waves. Spectroscopy was first used as a means of identifying chemical elements.

The range of light waves (electromagnetic spectrum) emitted by an element heated to incandescence consists of a set of bright lines at wavelengths unique to that particular element. When white light passes through a gas containing the same element, the spectrum produced has dark lines at exactly the same wavelengths as a spectrum produced by incandescence. In either case, as the amount of the element present increases, the intensity of the bright lines increases or the dark lines become darker.

SPECTROPHOTOMETRY: VISIBLE-LIGHT SPECTROSCOPY

Spectroscopy also works on liquid solutions of pure substances, which absorb electromagnetic radiation only at certain wavelengths. *Spectrophotometry* is a type of spectroscopy that relies on the absorption of electromagnetic radiation, at or near the range of visible light, at specific wavelengths through a certain substance, often a substance in solution. So, using spectrophotometry to analyze a solution requires a source of light that can be adjusted to a known wavelength and a means of measuring the intensity of the light after the light has passed through the solution.

A special type of test tube called a *cuvette* holds the solution being analyzed. Cuvettes have flat parallel sides, and transmit light very uniformly. This way, the conditions under which samples are analyzed can be as controlled as possible from one test to the next. On many spectrophotometers, you can set the meter to display the measured light intensity as either percent transmittance, which is a measure of how much light of the measured wavelength came through the solution, or as absorbance, which is the opposite: the amount of light that the solution absorbed, which did not pass through.

A Lesson on Spectroscopy *continued*

FIGURE 1 SPECTROPHOTOMETER SCHEMATIC

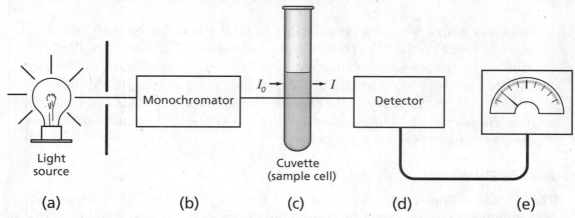

(a)   (b)   (c)   (d)   (e)

A basic schematic of a spectrophotometer is shown above in **Figure 1.** A light source, (a), emits light, which passes through a monochromator, (b). The monochromator filters the light so that only light of the desired frequency will pass through the sample chamber, (c), containing a cuvette with the solution to be tested. I_0 in **Figure 1** above represents the intensity of the light as it enters the sample. The intensity of the light, I, is measured by a detector, (d). The difference between I_0 and I is displayed on a meter, (e), as the absorbance, or transmittance, depending on what the machine has been set to measure.

Absorbance, A, is the logarithm of the ratio of the light intensity entering the sample, I_0, to the light intensity emerging from the sample, I.

$$A = \log (I_0/I)$$

So, the more light a substance lets through, the closer I will be to I_0 and the lower the absorbance value will be. Note that absorbance is a number without units and that it can have any positive value. The percent transmittance, %T, of a solution is 100% multiplied by the ratio of the light intensity emerging from the sample to the light intensity falling on the sample.

$$\%T = 100 \times (I/I_0)$$

The equation above reflects the fact that the more light that passes through the sample, the higher the percent transmittance will be. Absorbance can be found from percent transmittance using the equation below.

$$A = \log \left(\frac{100\%}{\%T} \right)$$

After setting a certain wavelength and before taking measurements, a spectrophotometer must be calibrated by placing a cuvette containing the solvent used to make the solution into the sample compartment. The spectrophotometer is then set to read zero absorbance (or 100% transmittance). This cancels out any absorption by the cuvette or by the solvent itself and controls for any variation in light intensity at different wavelengths. This calibration, or "zeroing out," is done to ensure that any light absorption measured by the spectrophotometer will be due only to the substance being measured.

A Lesson on Spectroscopy *continued*

FIGURE 2 ABSORPTION SPECTRUM OF A LEAF PIGMENT EXTRACT

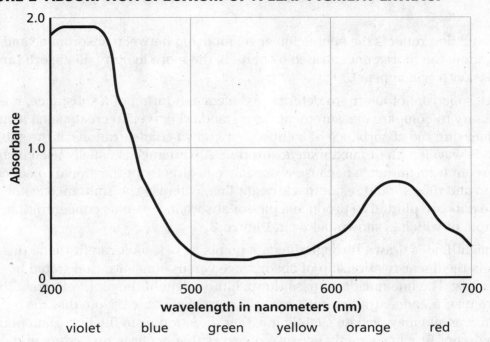

How to Use Spectrophotometry
IDENTIFYING A SUBSTANCE BY ABSORPTION CURVE

Spectrophotometry can be used to identify an unknown substance. If data is taken for absorbance readings over a range of wavelengths and plotted as a graph of absorbance versus wavelength, they produce an absorption curve that is characteristic of that substance. An example of such an absorption spectrum is shown in **Figure 2** above. An absorption curve can be plotted for an unknown substance and compared with that of a known substance suspected to be the same as the unknown substance. If their absorption curves match, they are probably the same. The height of the absorbance peaks may vary from sample to sample based on concentration; the unique identifying factor, however, is the wavelength at which these peaks occur in an absorption spectrum.

MEASURING CONCENTRATION BY USING BEER'S LAW

A spectrophotometer can also be used to measure concentration. The relationship between absorbance and concentration is *Beer's law*, which can be expressed as $A = abC$, where A is the measured absorbance of the solution, a is the *absorptivity* of the substance, b is the length of the light path through the substance, and C is the concentration of the substance in solution. Absorptivity, a, is a constant for a given substance, and b is a constant for a given spectrophotometer. This means that a and b can simply be disregarded during an experiment using the same substance in solution and the same spectrophotometer.

A Lesson on Spectroscopy *continued*

This leaves a simple proportionality between absorbance and concentration:

$$A \propto C.$$

This equation reflects the simple linear relationship between absorbance and concentration: the more concentrated a solution, the more light it will absorb (and the darker it will appear).

To use spectrophotometry to determine the concentration of a substance, it is necessary to compare measurements to a standard curve. To create such a curve, you measure the absorbance of solutions of known concentrations at the substance's wavelength of maximum absorption. Absorbance is usually used instead of percent transmittance because absorbance is directly proportional to concentration and thus should result in a straight line, whereas transmittance is not. These data are plotted to produce a plot of absorbance versus concentration, an example of which is shown below in **Figure 3.**

A straight line is drawn through the data points as best as it can fit them: this reflects the linear relationship of absorbance versus concentration stated by Beer's law. The line must also pass through the origin of the plot, reflecting the fact that zero concentration must result in zero absorbance. Note that the individual data points may deviate slightly in either direction from the line. Data points often do not fit a line exactly because of errors that occur in procedure and measurement. However, if a straight line fit is possible, the plot can be used to determine unknown concentrations. For most substances in solution, Beer's law is only valid for absorbance values up to about 1. Beyond that range, inaccuracies tend to invalidate Beer's law.

FIGURE 3 AN EXAMPLE OF A BEER'S LAW STANDARD GRAPH

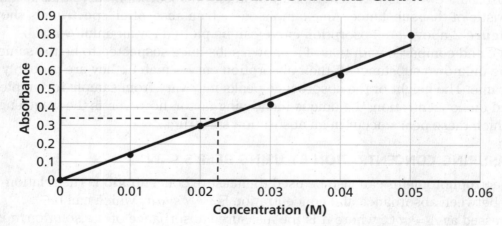

A Lesson on Spectroscopy *continued*

After a plot is done, the absorbance value of the unknown is determined using the same wavelength as the one used for the known concentrations. The concentration of the unknown can then be found by *interpolating* on the straight line in the standard plot. What this means is that after you take an absorbance reading of the unknown solution, you find where that absorbance occurs on the Beer's law plot. Then, you trace down to the x-axis from that point on the line to find the concentration that would be expected to provide that absorbance. For example, from the Beer's law graph in **Figure 3,** from an absorbance reading of 0.34, you would interpolate a concentration value of about 0.023 M.

IDENTIFYING A SUBSTANCE BY PARTICLE SETTLING RATE CURVE

A solid such as soil suspended in a liquid produces a condition called *turbidity:* this means that the liquid will appear cloudy. The cloudy appearance is the result of light reflected in random directions by the solid particles and light being absorbed by the solid particles. The amount of light that passes through such a liquid is a function of the amount of solid material suspended in the liquid. Spectrophotometry can therefore be used to measure absorbance in such a mixture.

Because much of the light passing into a turbid liquid is scattered rather than absorbed, the absorption spectrum produced is relatively flat with few, if any, peaks. An identifying absorbance curve, with peaks at various wavelengths, therefore cannot be obtained for a suspension as it can for a solution. Recall that a suspension is not a stable mixture: after having been mixed, it settles out upon standing. The turbidity of a suspension of a solid in liquid is related to its concentration of particles per liter, as well as how thoroughly it is mixed. Therefore, after you mix a suspension and measure its turbidity in a spectrophotometer, the rate of the mixture's settling can be measured by taking several absorbance readings over time.

As you have learned, density is a good way to identify a substance or compare two substances to see if they are the same. The speed at which the particulate substance in the water settles to the bottom of a container after being shaken up is also related to its density. Exact density determinations are hard to obtain in this way, but a settling rate curve can be used to compare two solids to see if they have the same density.

Topic Questions

1. If a substance has low absorbance, will its transmittance be low or high? Explain your answer.

Its transmittance will be high: absorbance, a measure of how much light is absorbed by a substance, is inversely related to transmittance, a measure of how much light is transmitted through a substance.

2. You have two solutions of copper sulfate, one a deep blue and the other much paler. Is the first or second likely to be more concentrated? Is the first or second likely to have higher absorbance?

The first solution is more concentrated. The first solution is also likely to have higher absorbance.

3. Two absorbance curves have peaks at the same wavelengths, but all of the absorbance values are greater on one of the two curves. Are the two curves plots of the same substance? Explain.

Yes, the curves are probably plots of the same substance because all of the absorbance peaks are at the same wavelengths. The difference in the absorbance values indicates that the concentrations of the samples that produced the curves are different. The curve with the smaller absorbance values was made from a sample with a lower concentration.

4. Why are absorbance values used instead of percent transmittance values when determining the concentration of a substance in solution?

Absorbance is directly proportional to concentration and will result in a standard plot that will be a straight line. Percent transmittance will not.

5. You have two soil samples, both of the same particle size distribution, and you want to compare them to see if they might have come from the same place. How might spectrophotometry be used to accomplish this?

Adding equal amounts of the soils to water, mixing them completely, and using a spectrophotometer to measure the rate at which the turbidity drops as the soil settles will allow you to compare the density of the two samples. If the densities match, then they are likely the same type of soil.

Practice Problems

1. The ratio of I_0 to I for a particular sample is 3:1. What would be the absorbance reading given by the sample?

$A = \log (I_0/I) = \log (3/1) = \log 3 = 0.477$

2. What would be the percent transmittance of the sample mentioned in item 1?

$\%T = 100 \times (I/I_0) = 100 \times (1/3) = 100 \times 0.333 = 33.3\%$

3. What would be the absorbance reading of a sample that had a percent transmittance of 80.%?

$A = \log \left(\dfrac{100\%}{\%T} \right) = \log (100\%/80.\%) = \log (1.25) = 0.097$

4. At a wavelength of 490 nm, one sample of a solution has an absorbance of 0.27 and a second sample has an absorbance of 0.81. The sample with the absorbance of 0.81 is found to have a molar concentration of 0.54 M. What is the molar concentration of the sample with the 0.27 absorbance?

An absorbance of 0.27 is one-third as great as an absorbance of 0.81.

Therefore, the molar concentration of the 0.27 absorbance sample is one-

third that of 0.54 M, which is 0.18 M.

The Fast-Food Arson (identification by spectroscopy absorbance peaks)

Teacher Notes

TIME REQUIRED one 50-minute class period

LAB RATINGS Easy ⟵ 1   2   3   4 ⟶ Hard
   Teacher Preparation–2
   Student Setup–1
   Concept Level–3
   Cleanup–1

SKILLS ACQUIRED
   Experimenting
   Inferring
   Interpreting
   Communicating

SCIENTIFIC METHODS

Make Observations Students will make observations and collect data in order to answer the experimental question.

Collect Data Students will collect experimental data.

Construct a Graph Students will plot a graph of their experimental data.

Analyze the Results Students will analyze their data in a systematic fashion.

Draw Conclusions Students will draw conclusions from their data in order to determine the answer to the experimental question.

Communicate the Results Students will clearly communicate their conclusions based on the outcome of the experiment.

MATERIALS (PER LAB GROUP)
- accelerant sample; from crime scene (1 drop green food coloring per 100 mL water)
- optical wipes, lint free
- paper, graph
- pipets (2)
- spectrophotometer, with accompanying cells
- water, distilled

SAFETY CAUTIONS

Students should wear safety goggles and aprons, and should use caution when handling sample cells. Notify janitorial staff when disposing of broken glass.

DISPOSAL

The food-coloring solutions can be disposed of down the drain. However, to help reinforce a sense of realism to the experiment, you may wish to have students dispose of the solutions in an aqueous chemical disposal container as if the solutions were really accelerant.

NOTES ON TECHNIQUE

Prior to the lab, discuss the use of the spectrophotometer with students. Directions for using the spectrophotometer are usually on the front of the machine. Students will be plotting an absorbance curve over a range of wavelengths; if the spectrophotometers you are using measure only transmittance you will have to show students how to convert transmittance values into absorbances (subtract the transmittance from 1 to obtain absorbance, and vice versa).

The two major uses of the spectrophotometer are to identify a material and to determine the concentration of a liquid. In this case, the identity of a material is being determined by comparing the absorbance curve of the unknown material to the absorbance curves of known materials.

You will need to test an aliquot of the food coloring solution with the spectrophotometer through a range of 450 nm to 600 nm to find out at what frequencies it gives the strongest absorbance readings (but that are less than 1.0). Make a note of these frequencies, and make a table like the one shown below:

Material	Absorbance characteristics
1	no absorbance peaks in the 500–650 nm range
2	a single absorbance peak at 420 nm
3	a single absorbance peak at 510 nm
4	absorbance peaks at 410 nm and 570 nm
5	absorbance peaks at 420 nm, 560 nm, and 640 nm

Replace the sample "absorbance characteristics" shown above with indicated peaks of your own choosing. If you want students' tests to make a positive match for the suspect substance, include in one of the entries the absorbance peak(s) you found the food coloring solution to give. Write the table on the board for students to refer to while doing their data analysis.

TIPS AND TRICKS

Because most schools do not have several spectrophotometers, you may want either to have more than two students per lab group or to arrange your lab schedule to allow time for all the lab groups to run all their tests. It may take several days to cycle all the lab groups through.

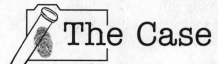

Forensics Lab

The Fast-Food Arson

The Case

Everyone in town's been talking about the big fire at Junior's. It was *the* place to go after football games and the softball tournament, so most folks had been there many times.

Not many people knew that Preston had been fired the night before by the manager. Those who did had overheard Preston mutter, "I'll get even for this," under his breath as he stormed out at closing time.

When the fire had been put out just before sunrise, the arson investigation team had brought Aladdin, their dog partner, to the scene. Aladdin's keen sense of smell swiftly found the spot where the fire started, in a storage area near the back door. A few milliliters of a liquid accelerant, which may have been used to start the fire, were recovered at the scene.

Now it's up to you, county forensic lab analyst. The arson team has brought you their sample, and a court-ordered search of Preston's garage has turned up five petroleum mixtures. Your job is to see if the sample from the scene matches any of the mixtures in the garage.

OBJECTIVE

Compare accelerant from a crime scene with samples from a suspect's garage using spectroscopy absorbance curves.

MATERIALS
- sample of crime scene accelerant
- distilled water
- graph paper
- optical wipes, lint free

EQUIPMENT
- pipets (2)
- spectrophotometer, with accompanying cells

SAFETY

- Always wear safety goggles and a lab apron to protect your eyes and clothing.

- Do not touch or taste any chemicals. Know the location of the emergency shower and eyewash station and know how to use them.

The Fast-Food Arson *continued*

• In the event a chemical gets on skin or clothing, wash the affected area immediately at the sink with copious amounts of water, keeping affected clothing away from skin. In the event of a chemical spill, notify the instructor immediately. Spills should be cleaned up promptly as directed by the instructor.

• In the event that you cut or puncture yourself with broken glass, notify your teacher immediately. Broken glass should be cleaned up as directed by your teacher.

Procedure

1. Create a data table in which to record absorbance values (ranging from 0.0 to 1.0) for wavelengths over the range 400 nm to 650 nm, in increments of 10 nm.

2. Follow the manufacturer's directions for the amount of time the spectrophotometer should warm up. Set the machine to display absorbance values. Fill a clean cell with distilled water using a pipet. Wipe any water from the outside of the cell using lint-free wipes. Place the cell in the sample compartment of the spectrophotometer.

3. Adjust the wavelength of the spectrophotometer to 400 nm. Adjust the calibration knob on the spectrophotometer until the absorbance reading for distilled water is zero. This way, the spectrophotometer will give a reading only for any substance that is added to pure water (in this case, the accelerant sample), so the water in the solution will not factor into the actual data you collect.

4. Remove the cell containing distilled water from the spectrophotometer, and place it to the side for the time being, as you will need it in a minute. Using a different pipet, fill a clean, empty cell with accelerant solution. Wipe any solution from the outside of the cell using lint-free wipes. Place the cell in the sample compartment of the spectrophotometer. Measure the absorbance value of the accelerant solution at 400 nm. Record this value in your data table.

5. Remove the cell containing the accelerant solution from the spectrophotometer. Increase the wavelength with which the spectrophotometer will measure the absorbance to 410 nm. Insert the cell with distilled water into the sample compartment. Adjust the calibration knob until the absorbance reading for distilled water is zero at 410 nm. Remove the cell and place the cell with accelerant in the sample compartment. Measure the absorbance of this solution at 410 nm, and record the absorbance value in your data table.

6. Repeat steps 3–5 over the range of 420 to 650 nm by increments of 10 nm. Each time you change the wavelength that the spectrometer reads, you must rezero the spectrophotometer with the cell containing distilled water prior to recording the absorbance value of the crime scene accelerant.

7. When all data collection is complete, dispose of all solutions as directed by your teacher.

The Fast-Food Arson *continued*

Analysis

1. Graphing Data After you have completed taking your data, plot the data on a piece of graph paper. Label the x-axis "Wavelength" and divide the axis into 10 nm increments from 400 nm to 650 nm. Label the *y*-axis "Absorbance" and mark it in equal intervals from zero to a convenient value slightly above your highest absorbance value. Then, plot the absorbance value for each wavelength on your graph and draw a smooth curve through the data points.

2. Analyzing Data Your teacher will provide you with a table giving the absorbance characteristics for the five suspicious petroleum mixtures from Preston's garage. Based on your results, do you think the accelerant obtained from the crime scene matches any of the substances found in Preston's garage? If so, indicate which, and what parts of your graphed absorbance data match the absorbance characteristics listed in the table given.

Results will vary based on the absorbance characteristics you provide.

Students should indicate a match of their data with one of the materials in

the table you provide if you gave that listed material absorbance characteris-

tics that matched the trial run you made on the solution prior to the lab.

3. Analyzing Methods Why is an established procedure so important in analysis of this type?

Without an established protocol that is repeated for each sample and wave-

length, the results and interpretation of the absorbance curve are open to

question.

Conclusions

1. Drawing Conclusions Should the police continue to investigate the possibility that Preston started the fire? What other investigations can be made to determine whether Preston started the fire?

If the accelerant sample does not match any of the samples found in

Preston's garage, police would have to look elsewhere for evidence connect-

ing Preston to the arson. If the accelerant sample matches, further tests

could be run on the chemical identities of the samples. Research should also

be done to determine if materials that yield absorbance peaks like that given

by the suspect material are common.

The Untimely Death (Beer's law interpolation using spectroscopy)

Teacher Notes

TIME REQUIRED two 50-minute periods

LAB RATINGS   Easy ◄—— 1   2   3   4 ——► Hard
  Teacher Preparation–3
  Student Setup–2
  Concept Level–4
  Cleanup–1

SKILLS ACQUIRED
  Experimenting
  Inferring
  Interpreting
  Communicating

SCIENTIFIC METHODS

Make Observations Students will make observations and collect data in order to answer the experimental question.

Collect Data Students will record data gathered from the experiment.

Construct a Graph Students will plot their data in a graph.

Analyze the Results Students will analyze their data in a systematic fashion.

Draw Conclusions Students will draw conclusions from their data in order to determine the answer to the experimental question.

Communicate the Results Students will clearly communicate their conclusions based on the outcome of the experiment.

MATERIALS (PER LAB GROUP)

- beakers, small (4)
- graduated cylinder, 10 mL (4)
- optical wipes, lint free
- paper, graph
- pencil, grease
- pipets (2)
- sample of "0.050 M anesthetic" (a drop of green food coloring in 100 mL water)

- sample of "suspect anesthetic" (green food coloring in water; concentration may vary: see note below)
- spectrophotometer, with accompanying cells
- water, distilled

A solution of green food coloring in water will give good absorption peaks, but its actual concentration will be difficult to determine. You can prepare a standard solution for each lab group by adding a drop of green food coloring in 100 mL of water, and assign to it an arbitrary concentration of "0.050 M." Lab groups will use graduated cylinders and beakers to make serial dilutions of this standard to concentrations of "0.025 M," "0.0125 M," and "0.00625 M."

Make another batch of standard solution in order to make unknown "suspect anesthetic" solutions by diluting. The safety limit of the anesthetic solution will be considered to be "0.020 M," so a suspect solution diluted at a ratio of less than (2 volumes of standard solution) : (3 volumes of distilled water) will yield a negative result, and less-dilute solutions should yield a positive test for lethal concentration of the "anesthetic."

SAFETY CAUTIONS

Students should work with caution when handling glassware and cells. As usual, appropriate protective goggles and aprons should be worn in the event an accident may produce flying projectiles. Notify janitorial staff when disposing of broken glass.

DISPOSAL

The food-coloring solutions can be disposed of down the drain. However, to help reinforce a sense of realism to the experiment, you may wish to have students dispose of the solutions in an aqueous chemical disposal container as if the solutions were really anesthetic.

NOTES ON TECHNIQUE

Prior to the lab, discuss the use of the spectrophotometer with students. The directions for using the spectrophotometer are usually on the front of the machine. Students will be plotting absorbance values for differing concentrations; if the spectrophotometers you are using measure only transmittance you will have to show students how to convert transmittance values into absorbances (subtract the transmittance from 1 to obtain absorbance, and vice versa).

The two major uses of a spectrophotometer are to identify a material and to determine the concentration of a liquid. In this case, the concentration of a material is being determined by plotting a Beer's law standard curve for four known concentrations. Beer's law states that the concentration of a solution will be directly proportional to absorbance as read by a spectrophotometer; thus, a standard absorbance curve of known concentrations will allow one to interpolate an unknown concentration from the absorbance. All absorbances must be between 0 and 1.0 for Beer's law to be valid.

You will need to test an aliquot of the "0.050 M" standard solution of food coloring with the spectrophotometer through a range of 450 nm to 600 nm to find out at what frequency it gives the strongest absorbance readings. Instruct students to take absorbance readings at the frequency at which you obtain the strongest absorbance reading.

Allow one class period for the preparation of the standard solutions and reading the absorbance values, and another period for the graphing and interpolation process.

If students mark the x-axis increments of their graphs of molar concentration as 0.00625, 0.0125, 0.025, and 0.050, the combined first three increments should be equal to the last increment, so make sure students mark 0.025 M halfway to 0.050 M. Students may be tempted to draw something other than a straight line through the origin and their four data points, but the graph must be linear in order to correctly interpolate and provide a Beer's law relationship. Instruct students that "connecting the dots" is of no use: it does not provide any information over and above the data points themselves. If there is a trend in a graph, it should be represented by a straight line interpolated among the points. Student plots must pass through the origin and show a straight line relationship between concentration and absorbance. The interpolation process for the unknown should be evident, and the concentration of the suspect anesthetic should be indicated on the x-axis.

TIPS AND TRICKS

Because students won't need to rezero the spectrophotometer more than once if they are careful, the experiment can be run relatively quickly, and one spectrophotometer should accommodate an entire class. Students can prepare their standard solutions at their lab stations and then simply insert their cells into the spectrophotometer, record the absorbance values, and return to their stations. Encourage the students to take great care in the preparation of their standard solutions because if the solutions are made carelessly, any interpolation using the graph would be inaccurate.

To make this activity a real challenge, you could make the suspect anesthetic solution have a concentration close to the "0.020 M" safety limit. Having the molarity of the suspect solution so close to the safety limit would give the accurate preparation of standard solutions even greater importance.

Name _____ Class _____ Date _____

Forensics Lab

The Untimely Death

The Case

When Darlene went to the hospital for knee surgery after an accident on the soft-ball field, her parents never suspected she wouldn't survive the operation.

The coroner has determined that Darlene had no conditions that could have caused her death during the surgery, so suspicion has shifted to the anesthesia administered. The anesthetic used during the operation may be safely used only in concentrations of 0.020 M or less. If it can be shown that the anesthetic was used at a concentration higher than this, then the anesthesiologist who gave Darlene the anesthetic would be liable for a malpractice suit.

Darlene's family has hired you, freelance chemical forensic investigator, to provide their legal team with hard evidence as to the cause of this tragedy. You have acquired a sample of the suspect anesthetic and will use a spectrophotometer to determine the concentration of anesthetic to see if it is greater than the safe amount. You will do so by interpolating from a graph of absorbance values of known concentrations.

OBJECTIVE

Determine the concentration of a solution using interpolation and Beer's Law.

MATERIALS
- sample of 0.050 M anesthetic
- sample of suspect anesthetic
- distilled water
- graph paper
- lint-free optical wipes
- grease pencil

EQUIPMENT
- spectrophotometer, with accompanying cells
- pipet (2)
- graduated cylinder, 10 mL
- small beakers (4)

SAFETY

- Always wear safety goggles and a lab apron to protect your eyes and clothing.

- Do not touch or taste any chemicals. Know the location of the emergency shower and eyewash station and know how to use them.

- If you get a chemical on your skin or clothing, wash it off at the sink while calling to the teacher. Notify the teacher of a spill. Spills should be cleaned up promptly, according to your teacher's directions.

- In the event that you cut or puncture yourself with broken glass, notify your teacher immediately. Broken glass should be cleaned up as directed by your teacher.

Procedure
MAKING STANDARD SOLUTIONS

1. Your teacher will provide you with a beaker of 0.050 M standard solution of the anesthetic. Label this beaker "0.050 M" with a grease pencil. Using a clean pipet, transfer 10 mL of this solution into one of the 10 mL graduated cylinders.

2. If you "overshoot" the 10 mL mark on the graduated cylinder, use the pipet to transfer solution out of the graduated cylinder, then pipet a smaller amount into the graduated cylinder, and repeat if necessary until you get as close as possible to the 10 mL mark. Accuracy is very important in making standard solutions.

3. Empty the 10 mL of solution into an empty beaker. Shake as many of the last remaining drops as you can into the beaker. Set aside the graduated cylinder you just used, and do not use it again for the experiment.

4. Take a clean 10 mL graduated cylinder and measure exactly 10 mL of distilled water into it as you did with the solution, taking care to get as close as possible to the 10 mL mark.

5. Empty as much as possible of the 10 mL of distilled water into the beaker containing the 10 mL of solution you measured. Swirl the beaker gently to mix as thoroughly as possible. Label this beaker "0.025 M."

6. Repeat from step 1, but this time take 10 mL of the solution you just diluted, and dilute it with 10 mL of distilled water as you did before (Use a different clean graduated cylinder for each dilution.). Label this dilution "0.0125 M." Repeat once more, diluting 10 mL of the 0.0125 M solution, which you just made, with another 10 mL of distilled water, and label this solution "0.00625 M."

READING ABSORBANCE VALUES

1. Create a data table in which to record absorbance values for concentrations of known anesthetic concentrations 0.050 M, 0.025 M, 0.0125 M, and 0.00625 M.

2. Follow the manufacturer's directions for the amount of time the spectrophotometer should warm up. Set the machine to display absorbance values. Fill a clean cell to the fill line with distilled water using a clean pipet. Wipe any water from the outside of the cell using lint-free optical wipes. Place the cell in the sample compartment of the spectrophotometer.

3. Set the machine to display absorbance values, adjust the wavelength to the wavelength your teacher specifies, and zero the machine so that it reads 0.00 A. This way, the spectrophotometer will give a reading only for any substance that is added to pure water (in this case, the anesthetic sample), so that the water in the solution will not factor into the actual data you collect.

4. Use a pipet to fill a clean cell to the fill line with the 0.050 M standard solution. Wipe the cell clean with a wipe, insert it into the spectrophotometer, and record in your data table the absorbance reading it gives.

5. Repeat step 4 with the 0.025 M, 0.0125 M, and 0.00625 M dilutions of the standard solution, recording in your data table each absorbance reading.

6. Obtain a sample of the suspect anesthetic of unknown concentration and find the absorbance value for this unknown in the same way as you did in steps 4 and 5.

7. When all data collection is complete, dispose of all solutions as directed by your teacher.

Analysis

1. **Graphing Data** The relationship between concentration and absorbance is known as Beer's law, which states that the more solute that is present in a solution, the more light will be absorbed by the solution, and also that this is a linear relationship. The first data point on your graph should be at a concentration of 0.00 M and absorbance of 0.00 A, because if the concentration of the solute in a solution is zero, then no light passing through the solution will be absorbed by the solute.

On a sheet of graph paper, label the *x*-axis "Concentration" and mark it in equal intervals from 0.00 M to 0.050 M. Label the *y*-axis "Absorbance" and mark it in equal intervals from 0.00 A to a convenient value slightly above your highest absorbance value (which should not be above 1).

Plot the concentration and absorbance values for your four standard solutions, and use a ruler to draw the best straight line through the four points and the origin of the graph. It is important that your line pass through the origin: zero concentration should result in zero absorbance. This is your *Beer's law standard graph*, which will allow you to relate absorbance to concentration in a solution of unknown concentration.

2. **Interpolating Data** Mark the absorbance of the suspect solution on the *y*-axis of your Beer's Law standard graph, and use a ruler to draw a horizontal line from the point where it intersects the *standard line*. Then use your ruler to draw a vertical line down from this intersection to the *x*-axis. This process is called *interpolation*. The point of intersection with the *x*-axis is your estimate of the concentration of the suspect anesthetic. Clearly indicate this concentration on your graph.

3. **Analyzing Methods** Why is preparation of the dilutions of the standard solution such an important part of an analysis like this one? Why would a defense attorney want to create doubt about how the standard solutions were made?

An accurate interpretation using the Beer's Law graph depends entirely on

how accurately the standard solutions were prepared. If a defense attorney

could create doubt as to the process of preparing the standards, then the

attorney could create doubt as to the accuracy of the interpolation.

4. **Analyzing Methods** For a Beer's Law interpolation, drawing a straight line through the data points is necessary because the Beer's Law relationship is a linear one. Why is drawing a straight line still acceptable even if the data points do not fit exactly on it?

Deviations of data points from a straight line are because of experimental

error such as imprecise solution preparation or inaccurate measurements

due to an imperfectly calibrated spectrophotometer. However, if the data

give a straight line fit in general, then a Beer's Law interpolation with the

data can be valid.

Conclusions

1. **Drawing Conclusions** Based on your results, do you believe the anesthesia was the cause of death in the patient? Explain.

Accurate results depend on accurate standard preparations, absorbance

value readings, and interpolations. If the suspect anesthesia has a concentra-

tion greater than 0.020 M, then there is a good chance that the anesthetic

was the cause of death, and therefore the anesthesiologist during the sur-

gery is responsible.

2. **Drawing Conclusions** What further tests or data analysis could be done in order to further eliminate doubts about the outcome of this procedure?

Possible answers include: repeating the experiment starting with a different

set of standard solutions, pooling the class data to get an average, and using

the least-square tool on a graphing calculator to get a better line fit among

the standard solution data points.

The Assault at the Flower Shop (soil-settling rate curve)

Teacher Notes

TIME REQUIRED several lab periods

LAB RATINGS   Easy ◀—— 1   2   3   4 ——▶ Hard
Teacher Preparation–2
Student Setup–2
Concept Level–3
Cleanup–1

SKILLS ACQUIRED
Designing experiments
Experimenting
Inferring
Interpreting
Communicating

SCIENTIFIC METHODS

Make Observations Students will make observations and collect data in order to answer the experimental question.

Collect Data Students will record data gathered from the experiment.

Construct a Graph Students will plot their data in a graph.

Analyze the Results Students will analyze their data in a systematic fashion.

Draw Conclusions Students will draw conclusions from their data in order to determine the answer to the experimental question.

Communicate the Results Students will clearly communicate their conclusions based on the outcome of the experiment.

MATERIALS (PER LAB GROUP)

- balances, digital
- graduated cylinder, 10 mL
- paper, weighing
- paper, graphing
- soil samples, from "Christine's clothing," "Jen's clothing," and "the flower shop"
- spectrophotometer, with accompanying cells
- water, distilled
- wire-mesh squares
- wipes, lint free

SAFETY CAUTIONS

Goggles and an apron should be worn at all times during this experiment. Students should exercise the usual care for handling glassware.

DISPOSAL

The soil samples can be discarded in the trash after use. Follow the spectrophotometer's instructions for cleaning the cells after use with soil samples.

NOTES ON TECHNIQUE

It is safer to use **sterilized** potting soils or soil mixes, because soil dug up from the outside probably contains mold and fungi to which students may be allergic.

Read your spectrophotometer instructions carefully to be sure it can handle soil samples before you attempt this lab.

If you want the outcome of the lab to identify Christine and Jen as guilty parties, use the same kind of soil for all three samples. Place the soil samples into plastic sandwich bags and mark as "Christine," "Jen," or "Crime scene." To prepare soil samples for testing, the students should pass some soil through wire mesh to remove debris and achieve uniform particle size. Wire mesh squares may be cut from window screen mesh, and then the edges should be covered with duct tape to prevent students from cutting or scratching themselves on the edges. A mass of 0.50 g of the meshed soil is about the right amount to conduct the tests. If the soils to be tested contain clay or other fine particles, the time required for settling will be too long for completion of the lab in one period. Assign student lab assistants the task of conducting sample settling curves. By analyzing these curves, you will obtain soil mixes that give differing curves and are manageable as far as time.

Prior to the lab, discuss the use of the spectrophotometer with your class. The directions for using the spectrophotometer are usually on the front of the machine. Because students will be plotting an absorbance vs. time graph, if the spectrophotometers you are using measure only transmittance you will have to show students how to convert transmittance values into absorbances (subtract the transmittance from 1 to obtain absorbance; and vice versa).

TIPS AND TRICKS

Because most schools do not have several spectrophotometers, your lab schedule should be arranged to allow time for all the lab groups to run all their tests. You will have multiple runs and resulting settling rate curves for each suspect and the crime scene. When everyone has conducted a run, allow time for the sharing of results. Some spectrophotometers have a computer interface feature, in which case the data points will be automatically taken and the graphs will be machine-generated, but it will still take several days to complete the analysis.

Name _____ Class _____ Date _____

The Assault at the Flower Shop

The Case

Staci has been attacked and robbed while returning to her car after closing her flower shop. As Staci left her shop after working quite late, she was attacked from behind by one person and wrestled to the ground. The day's cash receipts were stolen by a second person. In the course of the attack, Staci and her assailant ended up wrestling in an area where a fresh mix of special mulching soils had been dumped, and the composition of this unique soil may now be crucial to resolving the case.

Although Staci was unable to identify her attackers, she gave investigating officers the names of two former employees, Christine and Jen, who were recently fired by Staci and left their jobs vowing some form of revenge. They were located the following morning, and each denies any involvement in the attack and robbery. When questioned as to why some jeans in each of their laundry piles waiting to be washed had soil on the knees and seats, both Christine and Jen stated that they had been doing some yardwork the previous day. After the police looked at their yards, however, this claim seemed doubtful to the officers. Because the soil on Christine's and Jen's pairs of jeans is the only piece of evidence investigators have at this point, the soil needs to be analyzed to see if it is the same kind of soil as the mulching soil from the flower shop.

Your task as forensic lab analyst is to carefully analyze samples of the soils from each suspect's clothing and the flower shop mulching soil to see if a match can be made. You will gather soil settling rate data using a spectrophotometer and plot a curve. Your lab group will be responsible for testing only one of the three soils and will then compare your results with the work of others. Each type of soil will give a certain shape of curve, so comparing the curves will enable you to match soil types.

Objective

Plot a soil-settling rate curve of spectrophotometer data to determine if soil from suspect clothing matches that from a crime scene.

MATERIALS
- soil sample from either Christine's clothing, Jen's clothing, or the flower shop
- distilled water
- weighing paper
- graph paper
- wipes, lint free

The Assault at the Flower Shop *continued*

EQUIPMENT
- spectrophotometer, with accompanying cells
- digital balance
- graduated cylinder, 10 mL
- wire mesh

SAFETY

- Always wear safety goggles and a lab apron to protect your eyes and clothing.

- In the event that you cut or puncture yourself with broken glass, notify your teacher immediately. Broken glass and spilled soil should be cleaned up as directed by your teacher.

Procedure

Soil samples must be prepared for testing by passing them through wire mesh to remove debris and achieve uniform particle size among all the samples. This way, soil settling rates will be affected only by particle density—which will be the important factor in identifying the type of soil—and not particle size, which may differ throughout samples regardless of where they came from. Also, the masses of the soil samples tested must be equal. About 0.50 g is a good amount, but make sure whatever mass you use is the same for each sample.

After the spectrophotometer warms up, set the wavelength to 500 nm. Fill a spectrophotometer cell to the fill line with distilled water, and place it in the sample compartment. Use the calibrating knob to adjust the absorbance reading until it reads zero. This way, the absorbance of water itself will not be a factor in the measurements you make of the absorbance of soil samples.

Place your weighed soil sample into an empty spectrophotometer cell, and add distilled water to the fill line. Wipe the outside of the cell clean with a wipe. Just before inserting it into the spectrophotometer, cap the cell and vigorously shake it for two minutes. Allow any bubbles to rise to the top. To obtain a settling rate curve, take and record absorbance readings every one minute until five minutes after the absorbance value becomes stable. Initially the muddy water will cause most of the light to be absorbed, resulting in absorbance readings close to 1.00 or greater, but as the soil particles settle to the bottom of the test tube, the absorbance values will decrease. Each type of soil has a characteristic settling rate curve, and you are conducting this test to see if there is a match between either (or both) suspects and the soil known to have come from the floral shop crime scene.

1. Create a procedure that will allow you to plot the settling rate curve for the soil sample assigned to you. See the introduction pages on Spectroscopy, pp. 51–57, for hints. Discuss your procedure with your teacher for approval.

2. If your procedure is approved, carry out the experiment you have designed. Create a data table to record absorbance values and the time of each.

The Assault at the Flower Shop *continued*

Postlab Questions

1. Plot the absorbance values on the *y*-axis versus time (in minutes) on the *x*-axis. Discuss and compare your settling rate curve with lab groups analyzing the other suspect or crime scene samples.

2. Based on the class's total results, do you believe that the soil on either suspect's clothing came from the crime scene?

   Answers will vary. You may change the case outcome from class to class or

   year to year by placing crime scene soil into the plastic sandwich bags identi-

   fied as either suspect's soil sample.

3. Explain, in your own words, why it was necessary to sift the soil samples through wire mesh before testing them.

   Student answers should reflect the concept, stated above in the Procedure,

   that because particle size is likely to vary greatly both between and within

   soil samples it is not a reliable basis of comparison between soil types.

   Particle size is made uniform so that particle density will be the only factor

   that affects the results as reflected in the soil settling curves.

4. Why is an established procedure so important in analysis of this type?

   Without an established procedure, the results and interpretation of the tests

   are open to question. Soil particle size, mass of soil tested, water volume,

   shaking time, time between data points taken, absorbance wavelength, and

   x- and y-axis intervals should all be controlled; that is, they should not vary

   from sample to sample.

5. What other kinds of tests might be conducted on soil samples?

   Acidity, mineral content, moisture content, determination of types of

   admixed debris, pollen grains, or insect content are other characteristics

   that analysts could determine by testing.

The Assault at the Flower Shop *continued*

6. Generally, in what type of situation would soil type be of greatest value as evidence?

If soil types vary significantly throughout an area, soil analysis would help

narrow down the possible recent locations of a suspect. In contrast, if there

is very little variation in soil types in an area, this type of analysis is of

limited value. For instance, if all the soil in an area is sandy, it means very

little if a suspect has sandy soil on his or her clothing.

A Lesson on Identification

Suppose you are a police detective investigating the burglary of a jewelry store. The burglar must have worn gloves because no fingerprints were left. You have four likely suspects. How will you discover which person committed the crime? Solving this puzzle may be fairly easy if you can find a sample of the burglar's cells, such as blood, skin, or hair.

BACKGROUND

Biological evidence is highly prized because it is not subject to the same sources of error as other types of evidence about identity. Eyewitnesses often make mistakes. Physical evidence may show how the crime was committed, but it may not offer much information about who committed the crime.

Analysis of blood or DNA evidence, on the other hand, can give investigators very specific information that can positively identify or eliminate a suspect as having been at the scene of the crime. Biological evidence has its limitations, too. Contamination of samples can easily invalidate results, and laboratory specialists must be very careful in their analysis, as the samples they study are often very small. But ever since blood and DNA analysis methods have been developed, they have been important tools of identification for forensic scientists.

IDENTIFICATION BY BLOOD TYPE

Human blood can be one of four types—A, B, AB, or O—depending on which of two kinds of molecules, called *antigens*, are attached to the outer cell membrane of the red blood cells. These antigen molecules come in two forms: A antigens and B antigens. A person's blood is type A if only A antigens are present on the red blood cells. The blood is type B if it contains only B antigens. If the red blood cells of a blood sample have both kinds of antigens, the blood type is AB. If neither kind of antigen is present, the blood is type O. Because there are only four blood types, many people will share the same blood type, so analysis of blood type in certain cases may or may not be helpful in identifying or eliminating possible suspects.

IDENTIFICATION BY DNA

The nucleic acid known as DNA makes up every person's genetic information. Except in a case of identical twins (who have identical sets of DNA), each person's DNA is as unique as his or her fingerprints. The order of the *nucleotides*—the chemical units of information in DNA—determines what molecules are made in a person's cells. As a result, DNA determines most of the unique chemical and physical makeup of each person. A person's entire genetic code consists of a very long sequence of these base pairs, found in the nucleus in very compact bundles called *chromosomes*. **Figure 1** on the next page shows a diagram of DNA in chromosomes and the chromosomes in a cell's nucleus.

FIGURE 1 DNA, CHROMOSOMES, AND A CELL

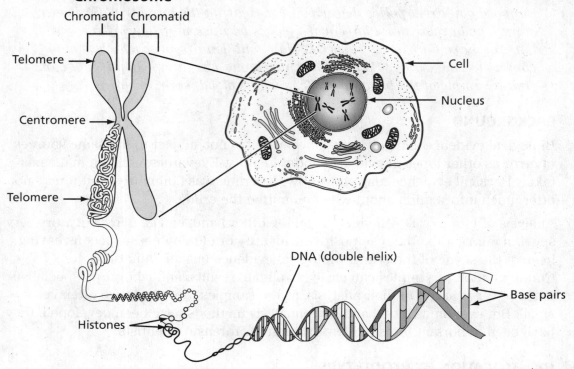

HOW IDENTIFICATION BY BLOOD TYPING WORKS

To determine blood type, you can treat a sample of blood of an unknown type with serum from type-A blood and then with serum from type-B blood. After each treatment, you must examine the red blood cells to see whether they clumped together. Clumping is a sign that the unknown blood contains antigens of a type opposite that of the added serum. For example, if the blood sample clumps when mixed with type-A serum, you know that the sample contains type-B antigens. The serum of type-A blood contains antibodies against the antigens of the cells in type-B blood. These antibodies, called *agglutinins*, attack the type-B red cells and cause them to clump together. Likewise, the serum from type-B blood contains antibodies that cause type-A red cells to clump. So, if type-A serum is added to a blood sample and clumping occurs, the sample may be type B.

However, because the sample may also be type AB, a test with B serum is needed. If the blood again forms clumps, it is type AB. If no clumps form, the sample is type B. Type-O blood will not form clumps with serum from type-A blood or with serum from type-B blood. Following a procedure that reflects the process shown in **Figure 2** below will allow you to determine, by antigen clumping, the type of an unknown sample of blood.

FIGURE 2 BLOOD-TYPING FLOWCHART

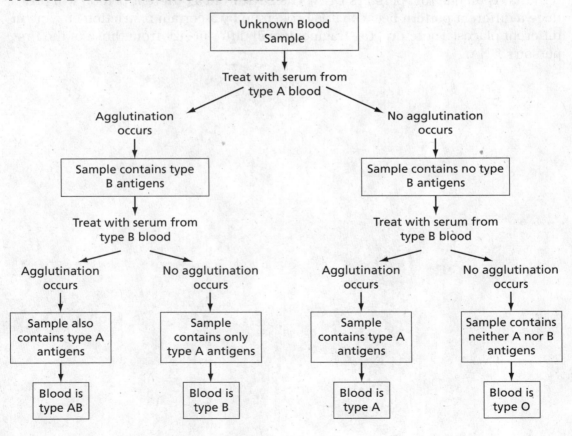

HOW IDENTIFICATION BY DNA WORKS

In forensic work, comparing the order of nucleotides in the DNA of a tissue sample with the DNA from a suspect can identify the sample, possibly determining the suspect's guilt or innocence. This procedure is called *DNA fingerprinting*. In DNA fingerprinting, patterns in the DNA from a crime scene are compared with patterns in the DNA of a suspect. If they match, investigators can be certain that the suspect was at the scene.

Given the extreme length of a DNA molecule, you may think that matching DNA from a crime scene with a suspect's DNA is very difficult. In fact, the process is fairly simple. Scientists have discovered enzymes from bacteria, called *restriction enzymes*, that break DNA chains only where certain nucleotide sequences occur. These nucleotide sequences are found in all human DNA, but at different places along the chain, depending on the person.

Suppose you break up a DNA sample with a restriction enzyme. The products will be DNA fragments of various sizes, depending on where the correct sequence of nucleotides occurred. These fragments can be separated according to size by a process called *electrophoresis*, in which DNA fragments migrate through a gel toward an electrode at various speeds according to their electrical charges (which are, in turn, determined by the size of the fragments). The result is a pattern that is unique to a person's DNA. The DNA from a different person will produce a different pattern because it is broken up by a certain restriction enzyme in different places. Therefore, the fragments will differ in size from those of the first person's DNA.

Topic Questions

1. What are the four main blood types?

Types A, B, AB, and O.

2. A student carrying out blood typing finds that the blood sample forms clumps when type-A serum is added. The sample forms no clumps when type-B serum is added. What is the type of the blood sample?

The sample is type B.

3. How does DNA determine most of the unique characteristics of a single person? Can a person have characteristics that are not determined by DNA? Explain and give two examples.

DNA forms the code for many of the molecules synthesized during life

processes. These molecules determine many of an individual's characteris-

tics. People have many characteristics unrelated to DNA. These characteris-

tics are acquired from the effects of the environment in which the person

lives. Examples could include scars, growth and development differences due

to diet or other environmental factors, diseases, differences in education,

behavioral differences, and life choices. Students may mention the differ-

ences that allow them to distinguish between identical twins.

4. What characteristics of DNA fragments determine the fragments' separation during electrophoresis?

Fragments have characteristic lengths based on their nucleotide sequences.

The fragments separate according to size (length of DNA chain).

5. Several different restriction enzymes exist. Suppose that you are trying to match the DNA from a crime scene with the DNA of a suspect. Why is it important to use the same restriction enzyme for both samples of DNA?

You must use the same restriction enzyme so that both samples of DNA, if

identical, will be broken in the same places, yielding fragments of the same

assortment of lengths. A second restriction enzyme may be used to carry out

a confirming test.

The Murder and the Blood Sample (identification by DNA analysis)

Teacher Notes

TIME REQUIRED One 50-minute period

RATINGS

Easy  Hard

Teacher Prep–2
Student Setup–2
Concept Level–4
Cleanup–2

SKILLS ACQUIRED

Collecting data
Communicating
Constructing models
Identifying patterns
Inferring
Interpreting
Analyzing data

SCIENTIFIC METHODS

Make Observations Students make observations as they model enzyme digestion and gel electrophoresis.

Analyze the Results Students analyze results in Analysis questions 1 and 2.

Draw Conclusions Students draw conclusions in Conclusions question 1.

MATERIALS (PER LAB GROUP)

- connectors, plastic (hydrogen bonds) (30)
- paper gel electrophoresis "lane"
- paper, legal size (8.5 in. × 14 in.)
- pop beads, blue (cytosine) (15)
- pop beads, green (guanine) (15)
- pop beads, orange (thymine) (15)
- pop beads, red (phosphate) (60)
- pop beads, white, 5-hole (deoxyribose) (60)
- pop beads, yellow (adenine) (15)
- restriction enzyme card Jan I
- restriction enzyme card Ward II
- ruler

Materials for this lab can be purchased as a kit from WARD'S.

Make copies of the Suspect/Victim DNA samples sheet. Cut apart the "Suspect" and "Victim" strips. Use a photocopy machine to make two additional strips from the DNA strip of the suspect that you choose to be the murderer and from the victim. Label these strips "crime scene sample 1" and "crime scene sample 2." For example, let's say you choose suspect 3 to be the murderer. Take a photocopy of the Suspect 3 DNA strip and cut out or blacken out the suspect's number and relabel as either "crime scene sample 1" or "crime scene sample 2." Do likewise with the "Victim" DNA strip. Assign one strip to each group in the class so that all strips are assigned at least once.

TIPS AND TRICKS

This lab works best in groups of two students.

Prepare an example of the gel electrophoresis page for students to see. You might want to prepare the gel electrophoresis pages for the entire class in advance.

The WARD'S kit has additional instructions on how to model the detection of the DNA through Southern blot analysis, which may be used to extend the scope of this lab.

The Murder and the Blood Sample *continued*

TEACHER'S GUIDE TO SUSPECT/VICTIM DNA AND RFLPS

Suspect 1 DNA

Suspect 2 DNA

Suspect 3 DNA

Suspect 4 DNA

Suspect 5 DNA

Victim's DNA

Key	Restriction Enzymes	Probe

77d

Forensics Lab

The Murder and the Blood Sample

The Case

The police are investigating a murder. Blood stains of two different types were found at the murder scene. Based on other forensic evidence, the police have reason to believe that the murderer was wounded at the time of the murder. The police currently have five suspects for the murder. You have been provided with the DNA from a blood sample of one of the five suspects, or the DNA from one of the two blood stains found at the crime scene.

BACKGROUND

In the early 1970s, scientists discovered that some bacteria have enzymes that are able to cut up DNA in a sequence-specific manner. These enzymes, now called *restriction enzymes*, recognize and bind to a specific short sequence of DNA, and then cut the DNA at specific sites within that sequence. Biologists found that they could use restriction enzymes to manipulate DNA. This ability formed the foundation for much of the biotechnology that exists today.

DNA fingerprinting is one important use of biotechnology. With the exception of identical twins, no two people have the same DNA sequence. Because each person has a DNA profile that is as unique as his or her fingerprints, DNA fingerprinting can be used to compare the DNA of different individuals.

In the first step of DNA fingerprinting, known and unknown samples are obtained and then digested, or cut into small fragments, by the same restriction enzyme. These short fragments are called *restriction fragment length polymorphisms (RFLPs)*. The next step in DNA fingerprinting is to separate the RFLPs by size. This is done with a technique called *gel electrophoresis*. The DNA is placed on a jellylike slab called a gel, and the gel is exposed to an electrical current. DNA has a negative electrical charge, so the RFLPs are attracted to the positive electrode when an electric current is applied. Shorter fragments travel faster and farther through the gel than longer ones. The length of a given DNA fragment can be determined by comparing its mobility on the gel with that of a sample containing DNA fragments of known sizes. The resulting pattern is unique for each individual.

In this lab, you will model experimental procedures involved in DNA fingerprinting and use your results to identify a hypothetical murderer.

OBJECTIVES

Use pop beads to model restriction enzyme digestion and agarose gel electrophoresis.

Evaluate the results of a model restriction enzyme digestion (DNA fingerprint).

Identify a hypothetical murderer by analyzing the simulated DNA fingerprints of suspects and DNA samples collected at the scene of the crime.

| The Murder and the Blood Sample *continued*

MATERIALS

- plastic connectors (hydrogen bonds) (30)
- paper gel electrophoresis "lane"
- paper, legal size (8.5 in. × 14 in.)
- pop beads, blue (cytosine) (15)
- pop beads, green (guanine) (15)
- pop beads, orange (thymine) (15)
- pop beads, red (phosphate) (60)
- pop beads, white, 5-hole (deoxyribose) (60)
- pop beads, yellow (adenine) (15)
- restriction enzyme card Jan I
- restriction enzyme card Ward II
- ruler

Procedure

1. Assemble the DNA you were assigned with pop beads, using the DNA strip given to your group as a blueprint. Use **Figure 1** to guide you in your assembly of your DNA "molecule." Be sure to assemble the beads in the precise pattern indicated, or your results will be incorrect. The assembled chain represents your subject's DNA.

FIGURE 1 PATTERN FOR ASSEMBLY OF POP BEADS

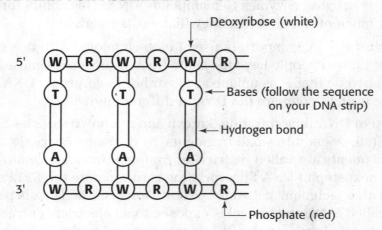

2. Place the DNA "molecule" you have just assembled on your work area so that the 5' end is on the top left side, as shown below. Be sure that the three orange beads (thymine) are in the following position on your work area:

5'   3'

TTT, etc....................G

AAA, etc.................C

3'   5'

Note: From this point on, it is important to keep the beads in this orientation. Do not allow the chain to be turned upside down or rotated. The 5' TTT end should always be on the top left of the molecule. If your chain is accidentally turned upside down, refer to your DNA strip to obtain the correct orientation.

3. Use the model restriction enzymes Jan I and Ward II to chop up the DNA. Look at your two restriction enzyme cards; they look like the cards in **Figure 2.** These "enzymes" will make cuts in the DNA in the manner indicated by the dotted lines.

FIGURE 2 RESTRICTION ENZYME CARDS

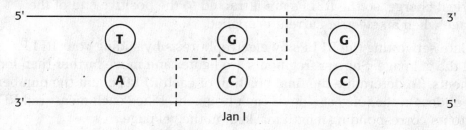

Jan I

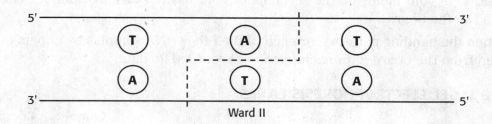

Ward II

4. Place restriction enzyme card Jan I on top of the left side of the DNA chain so that its label is right side up.

5. Move the card along the surface of the DNA until you match the precise sequence shown on the card. When you reach a sequence that matches the card, stop and break the beads apart in the manner indicated by the dotted lines.

6. Move the enzyme card until you reach the right end of the DNA. Double check the sequence with the enzyme card to ensure that you have made all the possible cuts.

7. Repeat the procedure on the remaining DNA fragments using the restriction enzyme card Ward II. Be sure to keep the DNA fragments in the orientation described above (5' orange thymine beads on the top left) throughout this exercise.

8. Create a gel electrophoresis area out of a legal size (8.5 in. × 14 in.) sheet of paper. On the left side of the paper, use a ruler to mark 1 in. increments from the bottom of the paper to the top of the paper. Starting at the bottom mark, label each mark from "0" (for the bottom mark) through "24" (for the top mark).

The Murder and the Blood Sample *continued*

9. Write a plus sign (+) at the bottom of the page and a minus sign (−) at the top of the page. Label the *y*-axis (left-hand margin) "Length of RFLPs (number of nucleotides."

10. Place the RFLPs at the negative pole of the gel electrophoresis page, taking care to retain the proper 5' to 3' orientation. Remember, DNA has a negative electrical charge, so the RFLPs are attracted to the positive end of the gel/page when an electric current is applied.

11. Simulate separating the RFLPs by electrophoresis by sliding your RFLPs along the gel/page. Shorter fragments are lighter and move farther than longer fragments. To determine the final position of each RFLP, count the number of nucleotides on the longest side of each fragment. Place each measured RFLP next to its corresponding length marked on the gel/page.

12. In the nine gel electrophoresis lanes in **Figure 3,** sketch dark bands at the correct positions in the gel lane reserved for your sample. Also, record the position of your bands on the seven lanes your teacher has provided for class data (on the board).

13. Obtain the banding patterns for each of the other DNA samples by copying them from the board after each team has recorded its data.

FIGURE 3 GEL ELECTROPHORESIS LANES

Victim's Blood	Suspect 1	Suspect 2	Suspect 3	Suspect 4	Suspect 5	Crime-Scene Sample 1	Crime-Scene Sample 2
22	22	22	22	22	22	22	22
20	20	20	20	20	20	20	20
18	18	18	18	18	18	18	18
16	16	16	16	16	16	16	16
14	14	14	14	14	14	14	14
12	12	12	12	12	12	12	12
10	10	10	10	10	10	10	10
8	8	8	8	8	8	8	8
6	6	6	6	6	6	6	6
4	4	4	4	4	4	4	4
2	2	2	2	2	2	2	2

The Murder and the Blood Sample *continued*

Analysis

1. Examining Data Are the RFLPs of the other DNA samples the same length as yours? Explain why or why not.

Answers will vary, but students should indicate that the lengths for most are

different because the DNA of each individual is unique. However, two sets of

samples will be the same; one DNA sample from the crime scene should

match the DNA from the victim's blood and the other DNA sample from the

crime scene should match the DNA of one of the suspects' blood.

2. Identifying Relationships Explain the role that restriction enzymes and gel electrophoresis play in DNA fingerprinting.

Restriction enzymes cut the DNA at a specific sequence, producing RFLPs.

Gel electrophoresis separates the RFLPs by size.

Conclusions

1. Drawing Conclusions Based on class data, which of the suspects is probably the murderer? Explain.

The identity of the murderer will vary. The suspect whose DNA sample

matches that of the blood left at the crime scene (that is not the victim's

blood) is probably the murderer.

2. Interpreting Information How did you show that the other sample found at the crime scene did not belong to the murderer?

The DNA fingerprint showed that the DNA from the other blood sample

found at the crime scene matched that of the victim.

3. Interpreting Information Imagine you are on a jury and that DNA fingerprinting evidence is introduced. Explain how you would regard such evidence.

Answers will vary. Accept all reasonable answers. In the case presented here,

the confirmation of the suspect's blood at the murder scene does not mean

that the suspect murdered the victim. It is only evidence that the suspect

was at the crime scene and left blood there. The DNA evidence must be

evaluated in conjunction with all of the other evidence before a conclusion

can be drawn.

Extensions

1. Research and Communications Look through newspapers and news magazines to find articles about actual court cases in which DNA fingerprinting was used to determine the innocence or guilt of a suspect in a crime. Share the articles with your classmates.

2. Research and Communications Do library research or search the Internet to find out more information about restriction enzymes and what role they play in bacteria.

Forensics Lab

Blood Typing (pre-laboratory exercise)

Teacher Notes

This experiment will serve as an introduction to blood typing and should be performed by the students prior to the students' attempting the subsequent "The Neighborhood Burglaries" lab.

TIME REQUIRED one 50-minute class period

LAB RATINGS

Teacher Preparation–4
Student Setup–2
Concept Level–3
Cleanup–2

Easy ←—— 1   2   3   4 ——→ Hard

SKILLS ACQUIRED

Experimenting
Measuring
Predicting
Organizing and Analyzing Data

SCIENTIFIC METHODS

Make Observations Students will make observations during the course of the experiment.

Analyze the Results Analysis question 1 requires students to analyze their results.

Draw Conclusions Conclusions questions 1–2 ask students to draw conclusions from their data.

MATERIALS (PER LAB GROUP)
• blood samples, of unknown type from four subjects, simulated
• serum, Anti-A, simulated (blood-typing)
• serum, Anti-B, simulated (blood-typing)
• serum, Anti-Rh, simulated (blood-typing)
• blood typing trays (4)
• toothpicks (12)
• wax pencil

Simulated blood samples can be purchased from many lab supply companies. Prepare these samples by relabeling them as needed. Because of blood-borne pathogens, actual body fluids should never be used in school labs.

SAFETY CAUTIONS

Have students wear safety goggles, gloves, and a lab apron.

DO NOT allow students to test any blood other than the simulated blood you provide.

Tell students that gloves provide protection from blood-borne pathogens and are necessary when working with body fluids.

DISPOSAL

Prepare separate containers for the disposal of broken plastic trays and toothpicks.

NOTES ON TECHNIQUE

Remind students to use a new stirrer (toothpick) for each test and to break used stirrers in two before placing them in the designated container.

Blood Typing

You are a new lab technician working in your city's blood bank. Before a person can donate blood, you must first determine his or her blood type. To make sure you know how to determine a person's blood type, your supervisor wants you to determine the blood type of four samples of simulated human blood.

BACKGROUND

Occasionally an injury or disorder is serious enough that a person must receive a *blood transfusion*, or blood from another person. A blood transfusion can succeed only if the blood of the *recipient*, the person receiving the blood, matches the blood of the *donor*, the person giving the blood. One of the factors that must be considered in matching blood is *blood type*. Blood type is determined by the presence or absence of specific marker proteins called *antigens* found on the surfaces of red blood cells. The most familiar blood typing system uses the letters *A*, *B*, and *O* to label the different antigens. Under this system, the primary blood types are A, B, AB, and O.

People with type A blood have a marker protein called the A antigen on the surface of their red blood cells. A person with type B blood has a slightly different marker protein called the B antigen. People with type AB blood have both A and B antigens on their red blood cells. People with type O blood have neither A nor B antigens on their red blood cells.

Antigen-antibody interactions are part of the immune system. Antigens label a cell as belonging to or not belonging to an organism. Antibodies produced in response to the presence of a foreign antigen attack the foreign antigen, defending the body against invasion.

Individuals produce antibodies, or *agglutinins*, to the marker proteins not found on their own cells. For example, individuals with type A blood produce antibodies against the B antigen, even if they have never been exposed to the antigen. Therefore, transfusions between people of different blood types usually are not successful because the recipient's antibodies will attack any blood cells that have marker proteins not found on the recipient's blood cells. **Table 1** on the next page summarizes the transfusion capabilities of the different blood types.

Name _____ Class _____ Date _____

Blood Typing continued

TABLE 1 TRANSFUSION CAPABILITIES OF DIFFERENT BLOOD TYPES

Blood type	Antigens on red blood cells	Antibodies in plasma	Can receive blood from groups	Can give blood to groups
A	A	B	O, A	A, AB
B	B	A	O, B	B, AB
AB	A and B	none	O, A, B, AB	AB
O	none	A and B	O	O, A, B, AB

If type A blood is transfused into a person with type B or type O blood, antibodies against the A antigen will attack the foreign red blood cells, causing them to clump together, or *agglutinate*. Clumps of red blood cells can block capillaries and cut off blood flow, which may be fatal. Clumping also occurs if type B blood is transfused into individuals with type A or type O blood, or if type AB blood is given to people with type O blood.

Blood typing is performed using *antiserum*, blood serum that contains specific antibodies. For ABO blood typing, antibodies against the A and B antigens are used. These antibodies are called anti-A and anti-B agglutinins. If agglutination occurs in the test blood only when it is exposed to anti-A serum, the blood contains the A antigen. The blood type of the test blood is A. If clumping occurs in the test blood only when it is exposed to anti-B serum, the blood contains the B antigen. The blood type of the test blood is B. If clotting occurs with both anti-A and anti-B sera, the type is AB, which has both A and B antigens. If no clumping occurs with either serum type, the blood type is O. This information is summarized in **Table 2** below.

TABLE 2 AGGLUTINATION REACTION OF ABO BLOOD-TYPING SERA

Reaction		Blood type
A antibodies (anti-A serum)	B antibodies (anti-B serum)	
agglutination	no agglutination	A
no agglutination	agglutination	B
agglutination	agglutination	AB
no agglutination	no agglutination	O

Another type of marker protein on the surface of red blood cells is the Rh factor, so named because it was originally identified in rhesus monkeys. People whose blood contains the Rh factor are said to be Rh positive (Rh^1). People whose blood does not contain the Rh factor are Rh negative (Rh^2). A person with Rh^2 blood has no antibodies to Rh^1 blood unless the person was exposed to Rh^1 blood at an earlier age. No agglutination occurs the first time an Rh^2 person receives a blood transfusion from an Rh^1 person. Agglutination can occur, however, the second time the Rh^2 person receives Rh^1 blood. In addition to testing for ABO blood type, it is also important to test blood for transfusion for its Rh factor.

OBJECTIVES

Determine the ABO and Rh blood types of unknown simulated blood samples.

MATERIALS
- simulated blood samples of unknown type from four subjects
- simulated Anti-A blood-typing serum
- simulated Anti-B blood-typing serum
- simulated Anti-Rh blood-typing serum
- toothpicks (12)

EQUIPMENT
- wax pencil
- blood typing trays (4)

SAFETY

- Always wear safety goggles and a lab apron to protect your eyes and clothing.
- Wear gloves at all times to protect your skin from the simulated blood.

Procedure

1. With a wax pencil, label each of four blood-typing trays as follows: "Tray 1—Mr. Thomas," "Tray 2—Ms. Chen," "Tray 3—Mr. Juarez," "Tray 4—Ms. Brown."

2. Place 3 to 4 drops of Mr. Thomas's simulated blood in each of the A, B, and Rh wells of Tray 1.

3. Place 3 to 4 drops of Ms. Chen's simulated blood in each of the A, B, and Rh wells of Tray 2.

4. Place 3 to 4 drops of Mr. Juarez's simulated blood in each of the A, B, and Rh wells of Tray 3.

5. Place 3 to 4 drops of Ms. Brown's simulated blood in each of the A, B, and Rh wells of Tray 4.

Blood Typing *continued*

6. Add 3 to 4 drops of the simulated anti-A serum to each A well on the four trays.

7. Add 3 to 4 drops of the simulated anti-B serum to each B well on the four trays.

8. Add 3 to 4 drops of the simulated anti-Rh serum to each Rh well on the four trays.

9. Use *separate* toothpicks to stir each sample of serum and blood. Record your observations in the data table below. Indicate an agglutination reaction with a "+" and no reaction with a "−." *Note: A positive test is indicated by obvious clumping of the red blood cells.* Also record your observations of each test.

10. Dispose of the toothpicks as directed by your teacher.

TABLE 3 BLOOD TYPING

	A antibodies with type A blood	B antibodies with type B blood	Rh antibodies with Rh+ blood	Blood type	Observation
Tray 1: Mr. Thomas	+	−	+	A+	Answers may vary.
Tray 2: Ms. Chen	−	+	−	B−	Answers may vary.
Tray 3: Mr. Juarez	+	+	+	AB+	Answers may vary.
Tray 4: Ms. Brown	−	−	−	O−	Answers may vary.

Analysis

1. **Applying Concepts** What factors determine the ABO blood types?

   The ABO blood typing is based on the presence or absence of antigens A and

   B, which are encoded by genes.

2. **Applying Concepts** What is the difference between an antigen and agglutinin?

   Antigens are substances found on the surface of red blood cells; agglutinins

   are antibodies found in plasma.

3. Applying Concepts If Ms. Brown were serving as a blood donor, what ABO blood type(s) could receive her blood safely?

Ms. Brown has type O– blood. Therefore, she can give blood to groups O–,

A–, B–, and AB.

4. Applying Concepts Which person among the four represented by the simulated blood samples can receive donated blood from Ms. Chen? Explain your answer.

Mr. Juarez can receive blood from Ms. Chen. Ms. Chen has blood type B–,

which can be received by Mr. Juarez, who has blood type AB+. A person with

type AB+ blood has A, B, and Rh antigens and therefore can receive blood

from a person with type A, B, AB, or O (whether Rh positive or negative).

Conclusions

1. Inferring Conclusions People with type O blood are commonly called universal donors. Explain why.

Type O blood carries no blood cell antigens to react with the recipient's

agglutinins. Type O blood can, therefore, be received by people with any

blood type.

2. Inferring Conclusions A person with what blood type would be considered a universal recipient? Explain your answer.

A person with type AB+ blood would be considered a universal recipient

because he or she can receive types A, B, AB, and O blood and either Rh+ or

Rh– blood.

Extension

1. Further Inquiry The first baby with Rh+ blood born to a woman with Rh– blood usually has no health problems. The second Rh+ child, however, can be seriously threatened before birth if the mother produces antibodies against the Rh antigens of her baby. Find out why this happens and what treatment is given to babies in this situation to save their lives.

2. Further Inquiry Find out what an emergency medical technician gives to a patient when the technician administers an emergency transfusion in the field. Why do technicians use this substance instead of blood for transfusing their patients?

The Neighborhood Burglaries (identification by blood typing)

Teacher Notes

TIME REQUIRED one 50-minute class period

LAB RATINGS Easy ←——1——2——3——4——→ Hard

Teacher Preparation–4
Student Setup–2
Concept Level–3
Cleanup–2

SKILLS ACQUIRED

Designing Experiments
Experimenting
Measuring
Predicting
Organizing and Analyzing Data

SCIENTIFIC METHODS

Ask Questions Students will ask a question to be answered by laboratory investigation.

Make Observations Students will make observations during the course of the experiment.

Analyze the Results Students will analyze and interpret their results.

Draw Conclusions Students will draw conclusions based on the lab objectives.

MATERIALS (PER LAB GROUP)

- blood sample, simulated, from "Victim"
- blood samples, simulated, for "Suspects" 1–4
- blood typing trays (6)
- compound microscope
- microscope slide (1)
- serum, Anti-A, simulated (blood-typing)
- serum, Anti-B, simulated (blood-typing)
- serum, Anti-Rh, simulated (blood-typing)
- toothpicks (18)
- water, distilled (10 mL)
- wax pencil

Simulated blood samples can be purchased from many lab supply companies. Prepare these samples by relabeling them as needed. Because of blood-borne pathogens, actual body fluids should never be used in school labs.

SAFETY CAUTIONS

Have students wear safety goggles, gloves, and a lab apron. DO NOT allow students to test any blood other than the simulated blood you provide. Tell students that gloves provide protection from blood-borne pathogens and are necessary when working with body fluids.

DISPOSAL

Stained cloth squares can be thrown away or washed and used again. Simulated blood and anti-serums can be washed down the drain or stored for future use. Prepare separate containers for the disposal of broken plastic trays and toothpicks.

NOTES ON TECHNIQUE

Students will narrow a list of suspects by determining the blood type of a simulated blood sample and comparing it with the types of five other samples. Prepare crime-scene samples by cutting 2 cm × 2 cm squares of cloth. Place them on a clean surface and dispense one drop of "crime scene" simulated blood onto each square, and allow them to dry. (To change the scenario from class to class, use a different suspect "blood" sample to stain the cloth squares.) Each lab group will need four cloth squares. Provide separate containers for collecting unused simulated blood, unused antiserums, used cloth squares, and broken glass.

Students should use one of the cloth samples to identify the substance as blood by looking for simulated red blood cells in the cloth fibers. Students should use the other three samples to determine the blood type. Adding water to these samples will rehydrate them to allow for blood typing.

Sample Student Response

QUESTION

Is the substance on the cloth blood, and if so, does it match the blood type of the victim or any of the four suspects?

PROPOSED PROCEDURE

1. Put on safety goggles, disposable gloves, and a lab apron.

2. Place one piece of stained cloth on a microscope slide, and place one drop of distilled water on the cloth. View the cloth under the microscope at low power and at high power to identify the stain. Record any observations, especially evidence that would prove that the substance in the stain is blood.

3. Using a wax pencil, label six blood-typing trays as follows: Tray 1—Crime scene; Tray 2—Victim; Tray 3—Suspect 1; Tray 4—Suspect 2; Tray 5—Suspect 3; and Tray 6—Suspect 4

4. Determine the type of blood in the cloth stains by placing a piece of the cloth in each of the A, B, and Rh wells of blood typing tray 1. Add several drops of water to each well to dampen the blood stains. Add 3 to 4 drops each of the anti-A, anti-B, and anti-Rh serums into the A, B, and Rh wells, respectively.

5. Using separate, clean toothpicks, stir the contents of each well in Tray 1 so that the serums mix with the stains. Record any observed agglutination in a data table.

6. Place 3 to 4 drops of the victim's blood into each of the A, B, and Rh wells of Tray 2. Place 3 to 4 drops of anti-A serum into the A well of the victim tray, add 3 to 4 drops of anti-B serum into the B well of the tray, and add 3 to 4 drops of anti-Rh serum into the Rh well of the tray. Using separate, clean toothpicks, stir the contents of each well.

7. Repeat step 6 for each of the suspects' blood samples.

8. In each well of each tray, look for evidence of agglutination and record observations in the data table. Use this data to determine the blood types of the stains, the victim, and each suspect.

9. Compare the blood type of each suspect with that of the sample found at the crime scene. Eliminate the suspects whose blood does not match the crime scene blood.

ESTIMATED COST OF THE JOB

$918.84, including a 30% profit (based on two students per group and one hour of labor)

RESULTS

Upon examining the crime-scene cloth sample under the microscope, students should observe simulated red blood cells that indicate the stain is indeed blood. In this case, the blood type of the crime scene stain is A+, a type that matches only that of suspect 3.

SAMPLE DATA TABLE

Blood source	Anti-A serum	Anti-B serum	Anti-Rh serum	Blood type
crime scene	agglutination	no agglutination	agglutination	A+
victim	no agglutination	agglutination	no agglutination	B–
suspect 1	agglutination	agglutination	agglutination	AB+
suspect 2	no agglutination	agglutination	no agglutination	B–
suspect 3	agglutination	no agglutination	agglutination	A+
suspect 4	no agglutination	no agglutination	no agglutination	O–

CONCLUSIONS

Students' data should support their conclusions. Students should find out that the stain on the cloth is indeed blood. They should suggest that, based on the results of the blood-typing evidence, if the blood stain came from the burglar, only suspect 3 could have committed the burglary.

Forensics Lab)   INQUIRY LAB

The Neighborhood Burglaries

Prerequisite

• "Blood Typing" lab on pages 83–87

 The Case

CITY OF OAKWOOD
POLICE DEPARTMENT
Oakwood, Missouri 65432-1221

December 18, 2005

Caitlin Noonan
Research and Development Division
BioLogical Resources, Inc.
101 Jonas Salk Dr.
Oakwood, MO 65432-1101

Dear Ms. Noonan,

We have recently had yet another series of burglaries. This time, all the victims have been residents of the same neighborhood. We were able to obtain some stained cloth samples from the last crime scene. According to the victim, he came home to find the burglar in his living room. The burglar broke a window to escape and was apparently cut by the glass. We believe that the cloth samples, which were found among the broken glass, are stained with the burglar's blood.

We are trying to solve this case as quickly as possible. Many of the residents in the area are concerned for their property and their safety. So far we have four suspects. All of them were seen near the crime scene, and all of them have cuts that could have been made by broken glass. We have blood samples from all four suspects and from the latest victim as well. Unfortunately, our forensics expert called in sick early this week and will not be able to complete blood-typing tests until next week. We need your research company to complete the necessary blood tests to help us narrow down our list of suspects. We will provide you with the stained cloth samples and the blood samples that we have collected. Please let me know what you find.

Sincerely,

Roberto Morales

Roberto Morales
Chief of Police
City of Oakwood Police Department

BioLogical Resources, Inc. Oakwood, MO 65432-1101

M E M O R A N D U M

To: Team Leader, Forensics Dept.

From: Caitlin Noonan, director of Research and Development

Chief Morales has sent the blood samples he described in his letter, along with a list of tests that need to be completed. Please have your research teams examine the stain on the cloth to determine whether it is indeed blood. Then complete a blood-typing test for the crime-scene stain samples, the victim, and each of the four suspects.

Chief Morales told me that he was very pleased with the work your team did on the last burglary case. I congratulate you on your previous success and encourage you to complete this project with the same exemplary skill and determination that you have demonstrated thus far.

Proposal Checklist

Before you start your work, you must submit a proposal for my approval. **Your proposal must include the following:**

_____ • the **question** you seek to answer

_____ • the **procedure** you will use

_____ • a detailed **data table** for recording results

_____ • a complete, itemized list of proposed **materials** and **costs** (including use of facilities, labor, and amounts needed)

Proposal Approval: _____

(Supervisor's signature)

The Neighborhood Burglaries *continued*

SAFETY ◈ ◈ ◈ ◈ ◈ ◈

- Always wear safety goggles and a lab apron to protect your eyes and clothing.
- Glassware is fragile. Notify the teacher of broken glass or cuts. Do not clean up broken glass or spills with broken glass unless the teacher tells you to do so.
- Never use electrical equipment around water or with wet hands or clothing. Never use equipment with frayed cords.
- Wash your hands before leaving the laboratory.
- Under no circumstances are you to test any blood other than the simulated blood samples provided by your teacher.

Procedure

When you finish your analysis, prepare a report in the form of a business letter to Chief Morales. **Your report must include the following:**

_____ • a paragraph describing the **procedure** you followed to examine the crime-scene blood sample and to complete blood-typing tests on all six blood samples.

_____ • a complete **data table** used for recording results

_____ • your **conclusions** about how the results indicate whether the crime-scene sample is indeed blood, and which, if any, of the other samples matches the blood type of the crime-scene sample

_____ • a completed **invoice** (see page 92) showing all materials, labor, and the total amount due

DISPOSAL

- Dispose of waste materials according to instructions from your teacher.
- Place used toothpicks in the waste disposal container indicated by your teacher.
- Place broken glass, unused simulated blood, unused antiserums, and used cloth squares in the separate containers provided.
- Wash reusable materials such as glassware and lab utensils, and return them to the supply area.

File: City of Oakwood Police Department

MATERIALS AND COSTS

(Select only what you will need. No refunds.)

I. Facilities and Equipment Use

Item	Rate	Number	Total
facilities	$480.00/day		
personal protective equipment	$10/day		
compound microscope	$30.00/day		
microscope slide with coverslip	$2.00/day		
scissors	$1.00/day		
ruler	$1.00/day		
hot plate	$15.00/day		
blood-typing tray	$5.00/day		
test tube	$2.00/day		
test-tube rack	$5.00/day		

II. Labor and Consumables

Item	Rate	Number	Total
labor	$40.00/hour		
4 stained cloth samples	provided		
vial of "Victim blood"	provided		
vial of "Suspect 1 blood"	provided		
vial of "Suspect 2 blood"	provided		
vial of "Suspect 3 blood"	provided		
vial of "Suspect 4 blood"	provided		
vial of anti-A typing serum	$20.00 each		
vial of anti-B typing serum	$20.00 each		
vial of anti-Rh typing serum	$20.00 each		
toothpicks	$0.10 each		
distilled water	$0.10/mL		
wax pencil	$2.00 each		

Fines

OSHA safety violation	$2,000.00/incident		
	Subtotal		
	Profit Margin		
	Total Amount Due		

Name _____ Class _____ Date _____

Topic Introduction

An Introduction to Environmental Chemistry

You've heard of "environmental" issues. Air pollution, water pollution, endangered animal species, and global warming may come to mind. The environment we live in—the air we breathe, the water we drink, and even the soil in which our food is grown—is subject to chemical changes that can affect health, climate, and many other important concerns. Consequently, there are many opportunities for chemistry to be applied in solving environmental problems.

Environmental chemistry is simply the study of the chemical makeup of the environment and chemical changes that take place in the environment. One area of environmental chemistry uses organisms such as bacteria and fungi to help solve environmental problems, such as cleaning up various kinds of chemical pollution.

A CURRENT HOT TOPIC: BIOREMEDIATION

"Remediation" means finding a remedy, or solution, for an existing problem; *bioremediation* is the use of biological materials to solve an existing problem. Problems such as groundwater contamination, chemical spills, and oil spills are a few examples in which the use of bioremediation is proving effective.

One type of bioremediation uses bacteria and fungi to remove substances from the environment as they grow. Often, they take in chemicals as food sources or as nutrients or trace substances that help them grow. In most cases, they chemically change a substance into final products that are safe and do not pollute the environment. One of the main advantages of bioremediation is that once the nutrients are used up, the organisms die and thus do not create additional pollution. Most of the organisms used for bioremediation are ones that already exist in the environment, so they are natural to the environment.

BIOAUGMENTATION

Most bioremediation involves *bioaugmentation*, which simply means adding organisms to the environment to help solve an environmental problem. Typically, one or more organisms are added to an area for the purpose of removing unwanted chemicals. Bacteria are the most commonly added organisms, but other organisms such as yeasts, algae, and fungi may also be used. Certain plant species have been found to remove toxic chemicals from soil and water. One plant example is locoweed, which can remove selenium from contaminated soils. In some cases, enzymes alone may be effective.

Certain kinds of chemicals are more *biodegradable* than others, as you probably know. This just means that some chemicals are easier to change into harmless chemicals than others are. This limitation applies to bioaugmentation, too: some chemicals cannot be easily changed by bioaugmentation into harmless ones. **Table 1** on the next page shows some classes of compounds with differing potentials for degradation.

Name _____ Class _____ Date _____

An Introduction to Environmental Chemistry *continued*

TABLE 1 BIODEGRADATION POTENTIAL FOR CLASSES OF COMPOUNDS

Organic compounds closer to the top of this table are more biodegradable than compounds closer to the bottom.

Compound class	Example	
Straight-chain hydrocarbon compounds	H H H H H H H H H–C–C–C–C–C–C–C–C–H H H H H H H H H *Octane*	**High potential**
Aromatic compounds	CH CH CH CH CH CH *Benzene*	
Chlorinated straight-chain compounds	H Cl C=C Cl Cl *Trichloroethylene (TCE)*	
Chlorinated aromatic compounds	X X X X X— —X X X X X X = H or Cl *Polychlorinated biphenyl (PCB)*	**Low potential**

BIOSTIMULATION AND BIOVENTING

Sometimes, the presence of certain naturally occurring organisms is a benefit to the environment. Research in the area of *biostimulation* is done to identify the particular nutrients that will encourage the growth of such organisms. One nutrient that is used in biostimulation is similar to ordinary fertilizer. Fertilizer supplies a source of nitrogen compounds, which are needed by most organisms but can be relatively scarce in many types of soil. Research has found that, in some cases, nutrients alone are not sufficient for some organisms' growth: they may also require a suitable surface, or substrate, to thrive, whereas others may need certain vitamins or minerals.

Microorganisms that are beneficial to the environment can also be helped by providing them with oxygen where they may otherwise have difficulty obtaining it. *Aerobic* organisms require oxygen to live, whereas *anaerobic* organisms are poisoned by the presence of oxygen. *Bioventing*, a variation of biostimulation, provides oxygen to aerobic organisms. Research in this area involves finding practical and economical methods of providing useful aerobic microorganisms with an abundant oxygen supply.

An Introduction to Environmental Chemistry *continued*

COMPOSTING AND BIOREACTORS

Gardeners make compost from weeds, plant trimmings, and grass clippings. These materials are allowed to "ferment" in a large pile, which causes the starting materials to break down. The resulting mixture of organic materials can be returned to the garden. Often, it will provide a good source of nutrients and trace elements, increasing plant growth.

Composting can also treat some materials contaminated with unwanted or hazardous chemicals. Composting includes a combination of several bioremediation techniques: biostimulation, bioventing, and bioaugmentation. Contaminated material is mixed with uncontaminated compost, bacteria, nutrients, and enough water to make the mixture slightly damp. The mixture is placed in a warm environment and aerated. After a period of time, the bacteria have converted the unwanted materials into harmless products. The composted mixture can then be used in a garden or placed in a landfill.

Bioreactors can perform some of the same tasks that composting accomplishes. A *bioreactor* is a large tank in which contaminated material is treated with microorganisms and enzymes. The materials in a bioreactor are usually kept mostly liquid. Bioreactors are used to remove pollutants from solid wastes, water, and soil. A diagram of a bioreactor being used to clean up contaminated groundwater is shown in **Figure 2** below. Treating large quantities of material with a bioreactor may be more expensive than other methods of bioremediation but may be the only alternative for some kinds of contamination.

FIGURE 2 A BIOREACTOR CLEANING UP CONTAMINATED GROUNDWATER

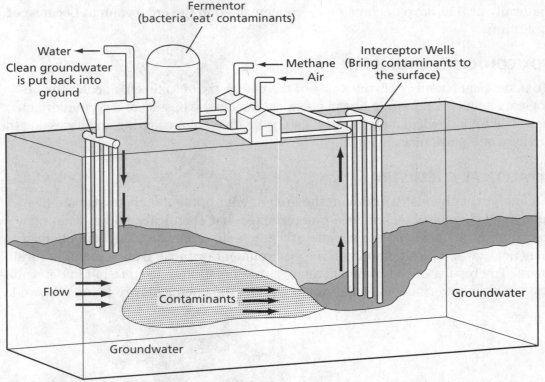

BIOFILTERS

A *biofilter* removes toxic or otherwise harmful materials from wastewater or gases before the water or gases are released into the environment. Wastewater treatment is probably the oldest example of the use of a biofilter. In a typical wastewater treatment, sewer water is passed slowly through a porous rock filter before it leaves the wastewater treatment facility. Bacteria growing on the rock break down various organic compounds contained in the wastewater. This process allows the output of a wastewater treatment plant to be safe enough to send out into the environment.

The treatment of organic gases, such as benzene vapor, is the main focus of current biofilter research. There are various types of filters that can be used, such as porous soil and compost, but all biofilters use microorganisms to break down the gas. A large part of this research involves finding organisms that will break down the gases. Once an organism is found, more work is done to find out how to keep it alive and effective.

Careers in Environmental Chemistry

In addition to the field of bioremediation, there are many other careers in chemistry that deal with environmental issues. Just a few are listed below.

ECOLOGICAL/ENVIRONMENTAL BIOCHEMISTRY

An ecological chemist studies the interaction between organisms and their environment and how chemicals in the environment affect these interactions. Of special interest to chemists in this field are chemicals that either do not occur naturally or that have become more concentrated in the environment because of pollution.

TOXICOLOGICAL CHEMISTRY

Toxicological chemistry is the study of the chemistry of toxic chemicals. Toxicological chemists may study how a chemical affects the tissues of a certain organism. The toxicological chemist may also be involved in measuring and assessing the toxicity of a particular chemical.

ANALYTICAL CHEMISTRY

An analytical chemist working in the area of environmental chemistry applies analytical chemistry techniques to determine what chemicals, and in what concentrations, are present in a sample of water, soil, or air. An analytical chemist may analyze samples taken from the environment or those from a manufacturing plant. The results of the chemical analysis are most often used to aid in determining whether government regulations regarding the chemicals are being followed.

An Introduction to Environmental Chemistry *continued*

Topic Questions

1. Describe in one sentence the purpose of bioremediation.

__Answers may vary. Sample answer: Bioremediation is used to remove__

__unwanted chemicals from the environment.__

2. What types of organisms can be used for bioremediation? What type of organism is most commonly used?

__Organisms used for bioremediation include yeasts, algae, and fungi. Bacteria__

__are the most commonly used organisms in bioremediation.__

3. What is the main advantage of using biological organisms to remove unwanted chemicals from the environment?

__Using biological organisms does not create any further pollution.__

4. How does composting differ from the use of a bioreactor?

__Composting is a relatively dry process, whereas a bioreactor is used with wet__

__materials. Bioreactors use enzymes and bacteria, whereas composting uses__

__only bacteria.__

5. A particular organism used for bioremediation is anaerobic. Would this organism be suitable for use in bioventing? Explain.

__No, it would not be appropriate. Because the organism is anaerobic, it will__

__die if it is exposed to the oxygen supplied by the bioventing process.__

6. Name and describe the branch of environmental chemistry that might be used to measure the concentration of a pollutant in a sample of air from the atmosphere.

__This would be a job for an environmental analytical chemist, who would use__

__analytical chemistry techniques on the sample to determine the concentra-__

__tion of the target substance.__

Extraction of Copper from Its Ore

Teacher Notes

TIME REQUIRED one 50-minute class period

LAB RATINGS

Easy ◄——— 1   2   3   4 ———► Hard

Teacher Preparation–3
Student Setup–2
Concept Level–2
Cleanup–3

SKILLS ACQUIRED

Constructing Models
Experimenting
Inferring

SCIENTIFIC METHODS

Make Observations Students will make observations during the experiment.

Analyze the Results Analysis questions 1 and 2 ask students to analyze their results.

Draw Conclusions Conclusions question 1 asks students to draw conclusions from their data.

MATERIALS (PER LAB GROUP)

- beaker, 500 mL
- copper(II) carbonate (about 15 g)
- iron filings (about 5 g)
- sulfuric acid, dilute (about 100 mL)
- Bunsen burner
- funnel
- test-tube holder
- test-tube rack
- test tubes, 13 mm × 100 mm (2)

Additional materials, such as a spatula and a stopwatch, may be helpful for students to use during the lab. Copper carbonate, $CuCO_3$, is more readily available and purer than malachite. Excellent results are obtained using $CuCO_3$.

SAFETY CAUTIONS

Make sure that students wear goggles, gloves, and lab aprons at all times. Before attempting this activity, become familiar with the material safety data sheet for sulfuric acid. When working with caustic or poisonous chemicals, use extreme caution and allow only your most mature students to handle these materials. A functioning eyewash station should also be immediately accessible. Because an open flame is being used in part of the lab, please address fire hazards, and review how to use fire extinguishers and fire blankets.

DISCUSSION

Before students begin this investigation, you may wish to discuss metallurgical processes. Metallurgy is the process whereby a metal is extracted from its ore and prepared for practical use. In their compounds, metals almost always exist in positive oxidation states. Therefore, the metal must be reduced (gain electrons) to extract it from its ore. Ores that contain impurities are treated to concentrate the metal and to convert some metal compounds into substances that can be more easily reduced.

NOTES ON TECHNIQUE

Have students add 100–150 mL of sulfuric acid (1 M) to the heated copper carbonate. When copper oxide is redissolved in sulfuric acid (step 6), the blue color indicates the presence of the hydrated copper(II) ion. There will probably be some copper oxide left in the test tube. If you wish to save it, you may direct students to wash it into a safe container.

Make sure students understand what they are observing in step 8. When the iron filings are added to the test tube containing copper sulfate, the copper ions are reduced to solid copper, which forms around the iron filings.

Environmental Chemistry Lab

Extraction of Copper from Its Ore

Most metals are combined with other elements in Earth's crust. A material in the crust that is a profitable source of an element is called an ore. Malachite (MAL uh KIET) is the basic carbonate of copper. The green corrosion that forms on copper is due to chemical weathering. *Chemical weathering* is the process by which material such as rocks breaks down as a result of chemical reactions. This green corrosion has the same composition that malachite does. The reactions of malachite are similar to those of copper carbonate.

In this investigation, you will extract copper from copper carbonate using heat and dilute sulfuric acid. The process you will be using will be similar to the process in which copper is extracted from malachite ore.

OBJECTIVES

Perform an extraction of copper from copper carbonate in much the same way that copper is extracted from malachite ore.

Hypothesize how this process can be applied to extract other metallic elements from ores.

MATERIALS

- copper(II) carbonate (about 15 g)
- iron filings (about 5 g)
- sulfuric acid, dilute (about 100 mL)

EQUIPMENT

- beaker, 500 mL
- Bunsen burner
- funnel
- test-tube holder
- test-tube rack
- test tubes, 13 mm × 100 mm (2)

SAFETY

- Put on a lab apron, safety goggles, and gloves.

- In this lab, you will be working with chemicals that can harm your skin and eyes or stain your skin and clothing. If you get a chemical on your skin or clothing, wash it off at the sink while calling to your teacher. If you get a chemical in your eyes, immediately flush it out at the eyewash station while calling to your teacher.

Procedure

1. **CAUTION: Wear your laboratory apron, gloves, and safety goggles throughout the investigation.** Fill one of the test tubes about one-fourth full of copper carbonate. Record the color of the copper carbonate.

2. Light the Bunsen burner, and adjust the flame.

3. Heat the copper carbonate by holding the test tube over the flame with a test-tube holder, as shown in the figure on the next page. **CAUTION: When heating a test tube, point it away from yourself and other students. To prevent the test tube from breaking, heat it slowly by gently moving the test tube over the flame.** As you heat the copper carbonate, observe any changes in color.

4. Continue heating the test tube over the flame for 5 min.

5. Allow the test tube to cool. Observe any change in the volume of the material in the test tube. Then, place the test tube in the test-tube rack. Insert a funnel in the test tube, and add dilute sulfuric acid until the test tube is three-fourths full. **CAUTION: Avoid touching the sides of the test tube, which may be hot. If any of the acid gets on your skin or clothing, rinse immediately with cool water and alert your teacher.**

6. Allow the test tube to stand until some of the substance at the bottom of the test tube dissolves. After the sulfuric acid has dissolved some of the solid substance, note the color of the solution.

7. Use the second test tube to add more sulfuric acid to the first test tube until the first test tube is nearly full. Allow the first test tube to stand until more of the substance at the bottom of the test tube dissolves. Pour this solution (copper sulfate) into the second test tube.

8. Add a small number of iron filings to the second test tube. Observe what happens.

9. Clean all of the laboratory equipment, and dispose of the sulfuric acid as directed by your teacher.

Analysis

1. **Explaining Events** Disregarding any condensed water on the test-tube walls, what substance formed in the first test tube? Write the balanced chemical equation of the reaction that occurred. Explain any change in the volume of the new substance relative to the volume of the copper carbonate.

   <u>**The new substance is copper oxide. The chemical equation is $CuCO_3 \rightarrow CuO$**</u>

   <u>**$+ CO_2$. Because CO_2 gas is released during the reaction, there is less solid**</u>

   <u>**material left in the test tube.**</u>

2. Explaining Events When the iron filings were added to the second test tube, what indicated that a chemical reaction was taking place? Explain any change to the iron filings. Explain any change in the solution.

Bubbles rose from the solution. The iron filings became red in color because

the copper sulfate was reduced to solid copper, which formed around the iron

filings. The solution became less blue because the copper sulfate that caused

the blue color was being reduced and it precipitated out of the solution.

Conclusions

1. Drawing Conclusions Why was sulfuric acid used to extract copper from copper carbonate? Write the balanced equation for the reaction that occurs.

Sulfuric acid was used to extract the copper from copper carbonate because

the chemical reaction that takes place forms water (H_2O) and carbon dioxide

(CO_2), releasing the copper from the copper carbonate. The balanced equa-

tion is $CuCO_3 + H_2SO_4 \rightarrow CuSO_4 + H_2O + CO_2$

Extension

1. Analyzing Data Suppose that a certain deposit of copper ore contains a minimum of 1 percent copper by mass and that copper sells for $0.30/kg. Approximately how much could you spend to mine and process the copper from 100 kg of copper ore and remain profitable?

Less than $0.30 can be spent on mining and processing because only 1 kg of

copper can be extracted from 100 kg of ore that is 1 percent copper, and the

selling price for 1 kg of copper is $0.30.

Effects of Acid Rain on Plants

Teacher Notes

TIME REQUIRED one 50-minute class period

LAB RATINGS

Easy ← 1   2   3   4 → Hard

Teacher Preparation–2
Student Setup–3
Concept Level–2
Cleanup–3

SKILLS ACQUIRED

Predicting
Constructing Models
Experimenting
Collecting Data
Organizing and Analyzing Data
Communicating

SCIENTIFIC METHODS

Make Observations Students will observe the effects of a reaction that produces sulfur dioxide.

Ask Questions Students will make predictions about the effects of acid precipitation on plant growth.

Analyze the Results Analysis questions 1 and 2 ask students to analyze their results.

Draw Conclusions Conclusions questions 1 and 2 ask students to draw conclusions from their data.

MATERIALS (PER LAB GROUP)

- beaker, 50 mL
- clear plastic bags, large (2)
- houseplants of the same type, potted (2)
- sodium nitrite, 2 g
- sulfuric acid, 1 M (2 mL)
- twist tie or tape

SAFETY CAUTIONS

Students should wear a lab apron, protective gloves, and safety goggles at all times during this investigation. Before attempting this activity, become familiar with the material safety data sheet for sulfur dioxide. When working with caustic or poisonous chemicals, use extreme caution. Allow only your most mature students to handle these materials. Alternatively, you may wish to handle the chemicals yourself or perform the procedure as a demonstration.

For this activity, it is essential that a functioning fume hood is used to safely remove the sulfur dioxide gas. A functioning eyewash station should be immediately accessible. Test the bags beforehand for possible leaks. Use only bags that are free of leaks. Sulfur dioxide is very poisonous. Keep students at least 5 m from the simulation for the duration of the reaction. Pour all leftover acid solution into a safe container, and neutralize it to a pH of 7 by adding a dilute base, drop by drop, before pouring the solution down the drain.

Environmental Chemistry Lab

Effects of Acid Rain on Plants

Acid precipitation is one of the effects of air pollution. When pollutants that contain nitrogen or sulfur react with water vapor in clouds, dilute acid forms. These acids fall to Earth as acid precipitation.

Often, acid precipitation does not occur in the same place where the pollutants are released. The acid precipitation usually falls some distance downwind—sometimes hundreds of kilometers away. Thus, the sites where pollutants that cause acid precipitation are released may not suffer the effects of acid precipitation.

Coal-burning power plants are one source of air pollution. These power plants release sulfur dioxide into the air. Sulfur dioxide reacts with the water vapor in air to produce acid that contains sulfur. This acid later falls to Earth as acid precipitation.

In this investigation, you will create a chemical reaction that produces sulfur dioxide. The same acids that result from coal-burning power plants will form. You will see the effects of acid precipitation on living things—in this case, plants.

OBJECTIVES

Perform a chemical reaction that produces sulfur dioxide, a component of acid precipitation.

Hypothesize the effects acids containing sulfur will have on plants.

MATERIALS

- clear plastic bags, large (2)
- houseplants of the same type, potted (2)
- sodium nitrite, 2 g
- sulfuric acid, 1 M (2 mL)

EQUIPMENT

- beaker, 50 mL
- twist tie or tape

SAFETY

- Put on a lab apron, safety goggles, and gloves.

- In this lab, you will be working with chemicals that can harm your skin and eyes or stain your skin and clothing. If you get a chemical on your skin or clothing, wash it off at the sink while calling to your teacher. If you get a chemical in your eyes, immediately flush it out at the eyewash station while calling to your teacher.

Procedure

1. Place 2 g of sodium nitrite in a beaker. Place a plant and the beaker inside a plastic bag. Do not seal the bag. **CAUTION: Steps 2–4 should be carried out only under a fume hood or outdoors.**

2. Carefully add 2 mL of a 1 M solution of sulfuric acid to the beaker. Immediately seal the bag tightly, and secure the bag with a twist tie or tape. **CAUTION: Because this reaction produces sulfur dioxide, a toxic gas, the bag should have no leaks. If a leak occurs, move away from the bag until the reaction is complete and the gas has dissipated.**

3. Seal the same type of plant in an identical bag that does not contain sodium nitrite or sulfuric acid.

4. After 10 minutes, cut both bags open while keeping the plants and bags under the fume hood. Stay at least 5 m from the bags as the sulfur dioxide gas dissipates. Keep the plants and bags under the fume hood.

5. Predict the effects of the experiment on each plant over the next few days. Record your predictions.

6. Observe both plants over the next three days. Record your observations in a data table like the one shown below.

SAMPLE DATA TABLE

Day	Control plant	Experimental plant
1	normal	normal
2	normal	slightly wilted
3	normal	very wilted

Analysis

1. **Analyzing Models** In what ways is this a realistic model of acid precipitation?

   All of the basic ingredients are the same: sulfur dioxide, water, and living

   plants.

2. **Analyzing Models** In what ways is this experiment not a realistic simulation of acid precipitation?

   The acidity is much greater in the simulation than in real life; the length of

   exposure is much shorter; little or no precipitation actually forms; the

   simulation takes place in a sealed environment; and in the simulation, the

   effects of exposure to acid occur more rapidly.

Conclusions

1. Examining Data How closely did your predictions about the effects of the experiment on each plant match your observations?

Answers may vary.

2. Drawing Conclusions What does this experiment suggest about the effects of acid precipitation on plants?

The acid precipitation damages the vegetation.

Extension

1. Analyzing Models Would you expect to see similar effects occur as rapidly, more rapidly, or less rapidly in the environment? Explain your answer.

Effects would occur less rapidly in real life because the acid would be much

less concentrated.

2. Building Models Acid precipitation is damaging to plants because the sulfur dioxide contained in the water vapor clogs the openings on the surfaces of plants and interferes with photosynthesis. What kind of a safe model would demonstrate the damaging effects of acid precipitation in the form of water vapor on plant photosynthesis? Would this model be a realistic simulation of acid precipitation?

Answers may vary.

Wetlands Acid Spill (acid-base titration)

Teacher Notes

TIME REQUIRED one 50-minute class period

LAB RATINGS

Easy ←—— 1   2   3   4 ——→ Hard

Teacher Preparation–2
Student Setup–3
Concept Level–2
Cleanup–3

SKILLS ACQUIRED

Collecting data
Experimenting
Organizing and analyzing data
Interpreting
Drawing real-world conclusions

SCIENTIFIC METHODS

Make Observations Students will collect pH and volume data using a calculator-interfaced pH sensor.

Analyze the Results In Analysis questions 1–4, students will analyze the data from their experiments and identify patterns.

Draw Conclusions In Conclusions question 1, students will interpret the data and relate their results to the real-world example given in the introduction.

MATERIALS (PER LAB GROUP)

- beaker, 250 mL
- buret, 50 mL
- HCl solution (10 mL) (see note below)
- LabPro® or CBL2 interface
- magnetic stirrer (if available)
- NaOH solution, ~0.1 M (60 mL)
- pipet, 10 mL
- pipet bulb or pump
- ring stand
- stirring bar
- TI graphing calculator
- utility clamps (2)
- Vernier pH sensor
- water, distilled

The 0.1 M NaOH solution can be prepared by adding 4.0 g of NaOH to make 1 L of solution.

Prepare your unknown HCl samples in the concentration range of 0.075 to 0.15 M. The 0.075 M HCl can be prepared by adding 6.2 mL of concentrated HCl per liter of solution. The 0.15 M HCl can be prepared by adding 12.5 mL of concentrated HCl per liter of solution.

SAFETY CAUTIONS

Sodium hydroxide is a corrosive solid and is known to cause skin burns upon contact. Wear gloves when working with this substance to prepare solutions. When added to water, this substance evolves much energy as heat. This substance is very dangerous to eyes, so wear face and eye protection when you use this substance.

Hydrochloric acid is highly toxic when ingested or inhaled. It is also known to be severely corrosive to skin and eyes. Wear gloves and eye protection when you use this substance. When using chemicals, students should wear aprons, gloves, and goggles.

Graphing Calculator and Sensors
TIPS AND TRICKS

- Students should have the DataMate program loaded on their graphing calculators. Refer to Appendix B of Vernier's *Chemistry with Calculators* for instructions.

- An alternate way of determining the precise equivalence point of the titration is to take the second derivative of the pH-volume data. When DataMate is transferred to the calculator, a small program called PHDERIVS will be copied into the calculator as well (PHDERIVS will not be loaded onto TI-73 and TI-83 calculators because of their memory capabilities. If you would like a copy of PHDERIVS for these calculators, visit the Vernier Software & Technology Web site at www.vernier.com). PHDERIVS allows you to view first and second derivative plots of pH-volume data. PHDERIVS is set up to analyze volume data in L1 and pH data in L2. To run the program, follow this procedure:

  a. After you have collected pH-volume data, leave DataMate by selecting QUIT from the main screen.

  b. Start the PHDERIVS program. Note: On a TI-83 Plus, PHDERIVS will be already loaded as a program (press [PRGM], not [APPS]). On a TI-86, 89, 92, or 92 Plus, the PHDERIVS program will be listed alphabetically below the DATAMATE programs.

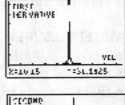

  c. Proceed to the GRAPHS menu. Select SECOND DERIV, a plot of $\Delta^2 pH/\Delta vol^2$.

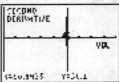

  d. Using the arrow keys, move the cursor to the equivalence point. This will be the point where the curve crosses the zero line. The x-value shown at the bottom of the screen is the volume of acid at the equivalence point.

TECHNIQUES TO DEMONSTRATE

Before they view graphs on the calculator, remind students how to use the arrow keys to trace the data points on the graph.

Experimental Setup

TIPS AND TRICKS

Consider dispensing the 10 mL of HCl required for each lab team from a buret instead of a pipet.

Lab teams should consist of two or three students. The titration will proceed faster if one student operates the buret while another student enters data on the calculator.

Consider having your students add two or three drops of phenolphthalein indicator at the beginning of each titration. They can then observe the phenolphthalein equivalence point and compare it with the pH equivalence point for the titration.

The pH calibrations that are stored in the DataMate data-collection program will work for this experiment. For more accurate pH readings, you (or your students) can do a 2-point calibration for each pH system using pH-4 and pH-7 buffers.

Answers

ANALYSIS

1. See data table.

2. See data table.

3. See data table.

4. See data table.

CONCLUSIONS

1. Based on the sample data below, 50 gal of 0.100 M NaOH would need to be added to neutralize the 50 gal of 0.100 M HC1.

DATA TABLES WITH SAMPLE DATA

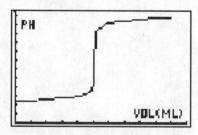

Acid-base titration using sodium hydroxide and hydrochloric acid

DATA TABLE

Concentration of NaOH	0.1000 M
Volume of NaOH added *before* largest pH change	9.80 mL
Volume of NaOH added *after* largest pH change	10.20 mL

Volume of NaOH added at equivalence point	10.00 mL
Mole NaOH	0.001 mol
Mole HCl	0.001 mol
Concentration of HCl	0.100 mol/L

Wetlands Acid Spill

An accident has recently occurred at a local wetlands refuge. A truck carrying drums of various chemicals overturned, and one of the drums ended up in a pond. On impact, the drum burst open, spilling its contents into the water. Markings on the drum indicate that it contained hydrochloric acid. Unfortunately, the markings that give its concentration are unreadable. Before cleanup crews can add a neutralizing agent to the water, they must first know the molar concentration of the HCl acid.

As an analyst for the EPA, you are responsible for determining the concentration of the unknown acid. A sample has been collected from the damaged drum. You will perform an acid-base titration to determine the moles of NaOH needed to neutralize the moles of HCl in the sample of acid collected.

The NaOH solution is of a known concentration and is the titration standard. Hydrogen ions from the HCl react with hydroxide ions from the NaOH in a one-to-one ratio to produce water in the overall reaction:

$$H^+(aq) + Cl^-(aq) + Na^+(aq) + OH^-(aq) \longrightarrow H_2O(l) + Na^+(aq) + Cl^-(aq)$$

In an acid-base titration involving HCl and NaOH, the initial pH of the acidic solution is very low. As the NaOH solution is added, the pH will change gradually. When the equivalence point is reached and all of the HCl has reacted, the pH change will be very rapid and the overall pH of the solution will go from acidic to basic. As additional NaOH is added, the pH will change gradually and level off.

A pH sensor will be used to monitor the pH of the solution during the titration. The volume of NaOH added to reach the equivalence point will be used to determine the molarity of the unknown HCl. The data collected will resemble the graph shown below.

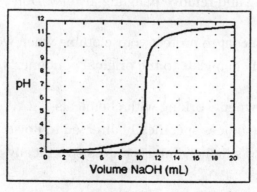

FIGURE 1

OBJECTIVE

Measure pH changes.

Graph pH-volume data pairs.

Identify the equivalence point of a titration curve.

Calculate the concentration of an unknown HC1 solution.

MATERIALS

- HCl solution, unknown concentration
- NaOH solution, ~0.1 M
- water, distilled

EQUIPMENT

- beaker, 250 mL
- buret, 50 mL
- LabPro or CBL2 interface
- magnetic stirrer (if available)
- pipet, 10 mL
- pipet bulb or pump
- ring stand
- stirring bar
- TI graphing calculator
- utility clamps (2)
- Vernier pH sensor

SAFETY

- Wear safety goggles when working around chemicals, acids, bases, flames, or heating devices. Contents under pressure may become projectiles and cause serious injury.

- Wear safety goggles when working around chemicals, acids, bases, flames, or heating devices. Contents under pressure may become projectiles and cause serious injury.

- Avoid wearing contact lenses in the lab.

- If any substance gets in your eyes, notify your instructor immediately, and flush your eyes with running water for at least 15 minutes.

- If a chemical is spilled on the floor or lab bench, alert your instructor, but do not clean it up yourself unless your teacher says it is OK to do so.

- Secure loose clothing and remove dangling jewelry. Don't wear open-toed shoes or sandals in the lab.

- Wear an apron or lab coat to protect your clothing when working with chemicals.

- Never return unused chemicals to the original container; follow instructions for proper disposal.

- Always use caution when working with chemicals.

- Never mix chemicals unless specifically directed to do so.

- Never taste, touch, or smell chemicals unless specifically directed to do so.

Procedure

EQUIPMENT PREPARATION

1. Obtain and wear goggles.

2. Pour 50 mL of distilled water into a 250 mL beaker. Use a pipet bulb (or pipet pump) to transfer 10 mL of the HC1 solution into the 250 mL beaker. **CAUTION: Handle the hydrochloric acid with care. It can cause painful burns if it comes in contact with the skin.**

3. Place the beaker on a magnetic stirrer and add a stirring bar. If no magnetic stirrer is available, you will need to stir the beaker with a stirring rod during the titration.

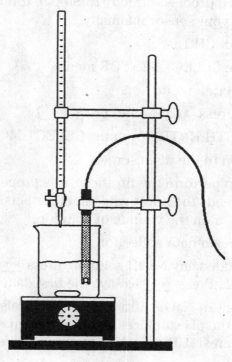

FIGURE 2

4. Plug the pH sensor into Channel 1 of the LabPro or CBL 2 interface. Use the link cable to connect the TI graphing calculator to the interface. Firmly press in the cable ends.

5. Use a utility clamp to suspend a pH sensor on a ring stand as shown in **Figure 2.** Position the pH sensor in the HCl solution, and adjust its position so that it is not struck by the stirring bar.

6. Obtain a 50 mL buret, and rinse with a few milliliters of the ~0.1 M NaOH solution. Dispose of the rinse solution as directed by your teacher. Use a utility clamp or buret clamp to attach the buret to the ring stand as shown in **Figure 2.** Fill the buret a little above the 0.00 mL level with ~0.1 M NaOH solution. Drain a small amount of NaOH solution so that it fills the buret tip *and* leaves the NaOH at the 0.00 mL level of the buret. Record the precise concentration of the NaOH solution in your data table. **CAUTION:** *Sodium hydroxide solution is caustic. Avoid spilling it on your skin or clothing.*

DATA COLLECTION

7. Turn on the calculator, and start the DATAMATE program. Press CLEAR to reset the program.

8. Set up the calculator and interface for the pH sensor.

   a. Select SETUP from the main screen.

   b. If CH 1 displays PH, proceed directly to Step 9. If it does not, continue with this step to set up your sensor manually.

   c. Press ENTER to select CH 1.

   d. Select PH from the SELECT SENSOR menu.

9. Set up the data-collection mode.

   a. To select MODE, press ▲ once and press ENTER.

   b. Select EVENTS WITH ENTRY from the SELECT MODE menu.

   c. Select OK to return to the main screen.

10. You are now ready to perform the titration. This process goes faster if one person adjusts and reads the buret while another person operates the calculator and enters the data on the volume of solution.

    a. Select START to begin data collection.

    b. Before you have added any NaOH solution, press ENTER and type in "0" as the buret volume in mL. Press ENTER to save the first data pair for this experiment.

    c. Add a small amount of NaOH titrant (enough to raise the pH about 0.15 units). When the pH stabilizes, press ENTER and enter the current buret reading (to the nearest 0.01 mL). You have now saved the second data pair for the experiment.

    d. Continue adding NaOH solution in amounts that raise the pH by about 0.15 units, and enter the volume reading from the buret each time. When a pH value of approximately 3.5 is reached, add and record only one drop at a time. Enter a new buret reading after each drop. (Note: It is important that all changes in volume in this part of the titration be equal; that is, one-drop increments.)

    e. After a pH value of approximately 10 is reached, again add larger increments that raise the pH by about 0.15 pH units, and enter the buret level after each increment.

    f. Continue adding NaOH solution until the pH value remains constant.

11. Press STO▶ when you have finished collecting data.

12. Examine the data on the displayed graph. As you move the cursor right or left on the graph, the volume (X) and pH (Y) values of each data point are displayed below the graph. Go to the region of the graph with the largest increase in pH. Find the NaOH volume just *before* this jump. Record this value in the data table. Then record the NaOH volume *after* the drop producing the largest pH increase was added.

13. Print a copy of the graph of pH versus volume.

14. (optional) Using the "Graphical Analysis" software, print a copy of the NaOH volume and pH data for the titration.

15. Dispose of the beaker contents as directed by your teacher. Rinse the pH sensor, and return it to the pH storage solution.

DATA TABLE

Concentration of NaOH		M
Volume of NaOH added *before* largest pH change		mL
Volume of NaOH added *after* largest pH change		mL

Volume of NaOH added at equivalence point		mL
Mole NaOH		mol
Mole HCl		mol
Concentration of HCl		mol/L

Analysis

1. Examining Data Use your data table to locate the equivalence point. The equivalence point is characterized as the region of greatest pH change. Determine the volume of NaOH added directly before and after the largest pH change. Add the two volumes together, and divide by two. Record the result in your data table as the volume of NaOH added at the equivalence point.

2. Organizing Data Using the volume of NaOH titrant added, calculate the number of moles of NaOH used to neutralize the HCl. Record the results in your data table.

3. Organizing Data Based on the equation for the neutralization reaction given in the introduction, calculate the number of moles of HCl used. Record the results in your data table.

4. Organizing Data Calculate the concentration of HCl used in this titration. Remember that you began with 10.0 mL of the unknown HCl solution. Record the results in your data table.

Conclusions

1. Drawing Conclusions The drum containing the HCl carried 189 L of acid. The entire contents of the drum were released into the pond at the accident site. According to the results of your titration, what concentration and volume

of NaOH would need to be added to neutralize the acid? _____

_____

_____

Environmental Chemistry Lab   **PROBEWARE LAB**

How Do Pollutants Affect a Lake?

Teacher Notes

TIME REQUIRED two 45-minute class periods, then 15 minutes every other day for five to seven days

LAB RATINGS   Easy ◄—— 1   2   3   4 ——► Hard
Teacher Preparation–3
Student Setup–3
Concept Level–2
Cleanup–2

SKILLS ACQUIRED
Collecting data   Interpreting
Experimenting   Measuring
Identifying patterns   Organizing and analyzing data
Inferring

SCIENTIFIC METHODS

Make Observations Students make observations during the experiment.

Form a Hypothesis Procedure step 2 asks students to form a hypothesis.

Test the Hypothesis Procedure steps 3–5 guide students in designing an experiment that will test their hypothesis. Students conduct an experiment that tests their hypothesis.

Analyze the Results Analysis questions 1 and 2 ask students to analyze their results.

Draw Conclusions Conclusions question 1 asks student to draw conclusions.

Communicate the Results Students communicate results in Analysis questions 1 and 2.

POSSIBLE MATERIALS (PER LAB GROUP)
- "lake water" containing several different species of algae (100 mL)
- CBL system
- DO calibration bottle
- DO probe
- DO electrode filling solution
- fertilizer (nitrate) solution (10 mL)
- fluorescent lights or grow lamp
- lab apron
- laundry detergent (phosphate) solution (10 mL)
- link cable
- pH probe
- plastic graduated pipets (3)
- rinse bottle of deionized water
- safety goggles
- sheet of white paper
- small jars or 50 mL beakers (3)
- TI graphing calculator
- wax pencil

Phosphate solution can be made with powdered laundry or dishwashing detergent or a powdered household cleaner (check label for phosphates and biodegradability) or with sodium phosphate, monobasic. Nitrate solution can be made with liquid or solid fertilizer containing no herbicides.

Prepare "lake water" by mixing any three or four of the following algae cultures: *Spirogyra, Chlorella, Chlamydomonas, Closterium, Zygnema, Oscillatoria,* and *Anabaena.* Use tap water that has been left out to dechlorinate for at least 24 hours.

Prepare detergent/fertilizer solutions by adding 12.5 g (e.g., powered detergent) or 12.5 mL (e.g., liquid fertilizer) and enough tap water to an Erlenmeyer flask to make 1 L. To break down the detergent and fertilizer so that nutrients are available to the algae, add about 5 g of dirt that is not sterile and that has no insecticides or herbicides added to it. Stopper the flask with cotton and leave it in a dark place for one week. Alternately, prepare phosphate solution by mixing 0.55 g sodium phosphate, monobasic with 99.45 mL distilled water. Filter the solutions before using them.

DISPOSAL

Adjust the pH of the pollution solutions to neutrality and sterilize the algae with bleach before washing down the drain with copious amounts of water.

NOTES ON TECHNIQUE

Demonstrate how to correctly handle probes (do not hit the bottom or sides of the beaker with the probe), and point out that fragile membrane at the tip of the dissolved oxygen (DO) probe.

TIPS AND TRICKS

This lab works best in groups of three to five students.

Refer to the DO probe booklet for instructions on polarizing and calibrating probes. The DO probes must be calibrated each day they are used. Provide students with the intercept and slope values you obtain. The DO probe must be polarized for 10 minutes. If more than one class is to perform this lab each day, students in earlier classes can leave their CBLs and calculators on when they finish. Students in the next class will then be able to skip all of the CBL setup, including the polarization.

This lab procedure calls for using the stored calibration values for pH probes. Recalibrating the pH probes might give more accurate results. To recalibrate, you can use two standard buffer solutions of pH 4 and pH 7.

How Do Pollutants Affect a Lake? *continued*

CHECKPOINTS

1. By the end of the first class period, have students turn in a detailed one-page plan/procedure for approval.

2. During the second class period, have students revise procedures according to the teacher's approval and begin the procedure.

3. After the first data collection, have students turn in a copy of their data so that the teacher can ensure they are collecting data correctly.

4. Following completion of the experiment, have students turn in their procedure, data, and Analysis and Conclusions questions. Students' labs should be evaluated on lab technique, quality and clarity of observations, and the explanation of observations and conclusions.

SAMPLE PROCEDURE

Prepare three containers with 27 mL each of "lake water." add 3 mL of fertilizer solution to one, 3 mL of phosphate solution to another, and 3 mL of water to the control. Place these preparations in an area where there is plenty of light for five to seven days. Use CBL probes to measure pH and/or dissolved oxygen on each day that class is held. Ideally, each group will test either the pH or dissolved oxygen of water containing one pollutant or the other. If different groups test different conditions, class data can be pooled to see the results for all conditions.

Name _____ Class _____ Date _____

How Do Pollutants Affect a Lake?

In this lab, you will design and conduct an experiment to determine how pollutants such as fertilizers and detergents affect the quality of water in a lake.

BACKGROUND

Lakes provide a home for a wide variety of organisms, including aquatic plants, fish, and a variety of arthropods, mollusks, and other invertebrates. The quality of the water in a lake affects the ability of these organisms to survive, grow, and reproduce. Aquatic organisms are sensitive to both the pH and the dissolved oxygen (DO) content of lake water. Organisms do best in lakes where the pH is between 6 and 9. A pH that is too high or too low can cause tissue damage and can increase the toxicity of compounds such as iron, ammonia, and mercury. Aquatic organisms are sensitive to the DO content of the lake water because they need oxygen to carry out cellular respiration. Cellular respiration provides these organisms with the energy they need to survive, grow, and reproduce.

As rainwater runs off agricultural and residential lands, it often carries pollutants, such as fertilizers, detergents, and fecal material from farm animals, into lakes. Pollutants can have many effects on a lake. Some pollutants are toxins, some change the pH of the lake, and some are actually rich sources of nutrients. Nitrates and phosphates, which are present in fertilizers and laundry detergents, are nutrients that are beneficial in small amounts for algae and plants.

However, when excess nutrients are present, a sudden massive growth of algae called an *algal bloom* may result. The development of an algal bloom in a lake often causes the death of many aquatic plants and animals.

1. What characteristics of lake water affect the health of aquatic organisms?

The pH of the lake and the amount of dissolved oxygen in the lake are

characteristics that affect the health of aquatic organisms. Accept other

reasonable answers.

2. How do pollutants such as nitrates and phosphates get into lake water?

Nitrates and phosphates are present in fertilizers, and phosphates are also

present in household detergents. As rainwater runs off agricultural and

residential land, it often becomes contaminated with these pollutants and

carries them into lakes.

3. Aquatic organisms require nitrates and phosphates to live. Under what circumstances do these nutrients become pollutants?

Nitrates and phosphates become pollutants when they are present in excessive

amounts. In excessive amounts, they cause the formation of an algal bloom.

Name _____ Class _____ Date _____

How Do Pollutants Affect a Lake? *continued*

SAFETY

- Wear safety goggles, gloves, and an apron at all times.
- Glassware is fragile. Notify the teacher of broken glass or cuts. Do not clean up broken glass or spills with broken glass unless the teacher tells you to do so.

OBJECTIVES

Develop a hypothesis about how common pollutants affect the quality of lake water.

Design and **conduct** an experiment to test your hypothesis.

Identify relationships between common pollutants and the pH and DO content of lake water.

Evaluate your results.

POSSIBLE MATERIALS

- "lake water" containing several different species of algae (100 mL)
- CBL system
- DO calibration bottle
- DO probe
- electrode filling solution
- fertilizer (nitrate) solution (10 mL)
- fluorescent lights or grow lamp
- lab apron
- laundry detergent (phosphate) solution (10 mL)
- link cable
- pH probe
- plastic graduated pipets (3)
- rinse bottle of deionized water
- safety goggles
- sheet of white paper
- small jars or 50 mL beakers (3)
- TI graphing calculator
- wax pencil

Procedure

FORMING A HYPOTHESIS

Based on what you have learned, form a hypothesis about how fertilizers and detergents might create an unhealthy environment for aquatic organisms.

1. What characteristics of the lake water might be changed by the presence of excess nitrates and/or phosphates?

Answers will vary. Students might suggest that excess nitrates and/or phos-

phates change any of the following: pH, DO content, amount of light avail-

able to organisms, or the amount of food available to organisms.

2. Write your own hypothesis. A possible hypothesis might be "The presence of excess nitrates changes the pH of the lake water to a level that is harmful to aquatic organisms."

Answers will vary. Hypotheses should be testable in an experiment using

materials from Possible Materials.

COMING UP WITH A PLAN

Plan and conduct an experiment that will determine what changes the pollutants in the lack cause that might be harmful to the organisms living there. Limit the number of conditions you choose for your experiment to those that can be completed during the time your teacher has allotted for this lab. Consult with your teacher to make sure that the conditions you have chosen are appropriate.

3. Write out a procedure for your experiment on a separate sheet of paper. As you plan the procedure, make the following decisions.
- Decide what pollutant(s) you will use.
- Decide what characteristics of the "lake water" you will observe or measure.
- Select the materials and technology that you will need for your experiment from those that your teacher has provided.
- Decide where you will conduct your experiment.
- Decide what your control(s) will be.
- Decide what safety procedures are necessary.

4. Using graph paper or a computer, construct tables to organize your data. Be sure your tables fit your investigation.

5. Have your teacher approve your plans.

PERFORMING THE EXPERIMENT

6. Put on safety goggles and a lab apron.

7. Implement your plan, using the equipment, technology, and safety procedures that you selected. Instructions for using CBL probes to measure pH and dissolved oxygen are included on the next page.

8. Record your observations and measurements in your tables. If necessary, revise your tables to include variables that you did not think of while planning your experiment.

9. When you have finished, clean and store your equipment. Recycle or dispose of all materials as instructed by your teacher.

Name _____ Class _____ Date _____

How Do Pollutants Affect a Lake? *continued*

SETTING UP AND USING THE PH PROBE

10. Plug the pH probe into the Channel 1 input of the CBL unit. Use the black cable to connect the CBL unit to the graphing calculator.

11. Turn on both the CBL unit and the calculator. Start the CHEMBIO program and go to the MAIN MENU.

12. Select SET UP PROBES. Enter "1" as the number of probes. Select pH from the SELECT PROBE menu. Enter "1" as the channel number.

13. Select USE STORED from the CALIBRATION menu.

14. Return to the MAIN MENU and select COLLECT DATA. Select MONITOR INPUT for the DATA COLLECTION menu. The CBL unit will display pH readings on the calculator.

15. Remove the pH probe from its storage solution. Use the rinse bottle filled with deionized water to carefully rinse the probe, catching the rinse water in a 500 mL beaker.

16. Submerge the pH probe in your sample of "lake water." When the pH reading stabilizes, record the pH in your table. Rinse the pH probe with deionized water between each reading.

17. After the final reading, rinse the pH probe with deionized water and return the probe to its storage solution. Dispose of the rinse water as instructed by your teacher. Press "+" on the calculator.

SETTING UP AND USING THE DISSOLVED OXYGEN (DO) PROBE

18. Plug the DO probe into the Channel 1 input of the CBL unit. Use the black cable to connect the CBL unit to the graphing calculator.

19. Turn on both the CBL unit and the calculator. Start the CHEMBIO program and go the the MAIN MENU.

20. Select SET UP PROBES. Enter "1" as the number of probes. Select D.OXY-GEN from the SELECT PROBE menu. Enter "1" as the channel number.

21. Select POLARIZE PROBE. Press ENTER to return to the CALIBRATION menu. You must allow the DO probe to polarize for 10 minutes before you can use it.

22. Select MANUAL ENTRY from the CALIBRATION menu. Enter the intercept (KO) and slope (K1) values for the DO calibration provided by your teacher.

23. After 10 minutes have passed, remove the DO probe from its storage solution. Submerge the probe in your sample of "lake water."

24. Select COLLECT DATA from the MAIN MENU. Select MONITOR INPUT from the DATA COLLECTION menu. Press ENTER.

25. Gently move the probe up and down about 1 cm in the sample. Be careful not to agitate the water, which will cause oxygen from the atmosphere to mix into the water. Continue moving the probe until the DO reading stabilizes. Record the DO concentration in your table.

26. Repeat steps 23 and 24 for each sample. Rinse the probe with deionized water between each reading.

27. After the final reading, rinse the DO probe with deionized water and return the probe to its storage solution. Press "+" on the calculator. Dispose of the rinse water as instructed by your teacher.

SETTING UP THE CBL SYSTEM FOR BOTH PROBES

To use both the pH probe and the DO probe, replace steps 12 and 20 with step 29, and replace steps 13 and 21 with steps 31 and 32.

28. Plug the pH probe into the Channel 1 input of the CBL unit. Plug the DO probe into the Channel 2 input. Use the black cable to connect the CBL unit to the graphing calculator.

29. Select SET UP PROBES. Enter "2" as the number of probes. Select pH from the SELECT PROBE menu. Enter "1" as the channel number.

30. Select MORE PROBES from the SELECT PROBE menu. Select D.OXYGEN from the SELECT PROBE menu. Enter "2" as the channel number.

31. Select POLARIZE PROBE. A message will appear. Select MANUAL ENTRY from the CALIBRATION menu. Enter the intercept (K0) and slope (K1) values for the dissolved oxygen calibration provided by your teacher. A message will appear concerning the sensors. Press ENTER. Leave the dissolved oxygen probe connected to the CBL for 10 minutes so that the probe can polarize.

32. Select COLLECT DATA from the MAIN MENU. Select MONITOR INPUT. Select either CH1 or CH2 from the SELECT A CHANNEL menu to monitor the probe reading. Use the CH VIEW button on the CBL to switch channels. Press TRIGGER on the CBL to quit monitoring. To view the other channel, select it from the SELECT A CHANNEL menu. To quit, choose QUIT from the SELECT A CHANNEL menu.

Analysis

1. Summarizing Data Summarize your findings and observations.

Answers will depend on students' hypotheses and choices. The addition of

nitrates to the "lake water" will likely cause an algal bloom. The DO content

of the "lake water" decreases due to cellular respiration and decomposition

of organisms in the algal bloom. The addition of nitrates and/or phosphates

may change the pH of the water. The pH may decrease as the algal bloom

releases carbon dioxide into the water, forming carbonic acid.

2. Describing Events Share your results with your classmates. Which hypotheses were supported?

Answers will vary. See answers to Analysis question 1.

3. Identifying Relationships How might an algal bloom contribute to a decrease in dissolved oxygen in the "lake water"?

Answers will vary, but students should support their answers with logical

rationales. Students might suggest the following: Algae use more oxygen

during cellular respiration than they make during photosynthesis. The

decomposition of dead algae uses up oxygen. The layer of algae prevents

other photosynthetic organisms from getting enough light to carry out

photosynthesis.

Conclusions

1. Drawing Conclusions What conclusions can you draw from your results? from class results?

Answers will vary, but students should recognize that the presence of the

pollutants results in a decrease in water quality that may be harmful to

many organisms.

2. Evaluating Models Was your experiment a good model for how pollutants might affect lake water? Explain why or why not, and give examples of what might be missing from your model.

Answers will vary, but students should support their answers. Students

might suggest that the effects of animals, plants, wind, rain, and soil are

missing in their model.

Evaluating Fuels (calorimetry)

Teacher Notes

TIME REQUIRED one 45-minute class period

LAB RATINGS Easy ←—1——2——3——4—→ Hard

    Teacher Preparation–2
    Student Setup–2
    Concept Level–3
    Cleanup–1

SKILLS ACQUIRED

    Collecting data
    Experimenting
    Interpreting
    Organizing and analyzing data

SCIENTIFIC METHODS

- **Analyze the Results** Analysis questions 6 and 7 ask students to analyze their results.

- **Draw Conclusions** Conclusions questions 3 and 4 ask students to draw conclusions.

MATERIALS (PER LAB GROUP)

- balance
- CBL system
- clamps, utility (2)
- crucibles (1–4)
- flame lighter
- flasks, 250 mL (1–4)
- fuels (1–4)

- lab apron
- link cable
- oven mitt or tongs
- ring stand
- safety goggles
- temperature probe
- TI graphing calculator

SAFETY CAUTIONS

Discuss all safety symbols and caution statements with students. Burn the fuels in a well-ventilated area or under a fume hood. Before the lab, review with students the procedures for handling flammable and combustible materials. Review the fire escape routes from the classroom. Do not allow students to use explosive fuels that have very low flash points, such as gasoline or white gas camping stove fuel. Be sure students position the probe cable so that it has no direct contact with the crucible. If alcohol burners are used, discuss their safe use.

Evaluating Fuels *continued*

DISPOSAL

When students have found the masses of the burned fuel samples, have them dispose of the samples in a large beaker or small metal bucket containing several centimeters of water provided for that purpose. After all burned sample materials are thoroughly extinguished by the water in the container, they may be discarded into a waste receptacle or otherwise disposed of in accordance with local or state regulations.

NOTES ON TECHNIQUE

Before the lab, help students determine which fuel or fuels they will investigate. Students might select twigs or different types of trees, such as oak, maple, or pine. Other possible fuel sources to test are paper, leaves, craft sticks, coal, a candle, or alcohol in an alcohol burner. An alcohol burner will not need to be placed into a crucible; the burner replaces the crucible. However, the height of the flask might need to be adjusted to be 2–3 cm above the flame.

Instead of having each group investigate several fuels, you might have different groups investigate different fuels and share results.

Demonstrate how to place and arrange fuel material in the crucible so that it is not overflowing and so that air space exists between individual pieces of fuel.

Before students light their first fuel source, check to see that they have followed Procedure step 6 correctly. Make sure the temperature probe wire does not come in contact with heat sources in the crucible, which could cause the wire insulation to melt.

Remind students not to stir the water with the temperature probe.

In Procedure step 8, remind students not to overfill the crucible with fuel or to pack down the fuel. Overfilling the crucible may result in pieces of the fuel sample falling out, invalidating the trial. Packing down the fuel might limit the amount of oxygen that can reach it.

TIPS AND TRICKS

This lab works best in groups of two to four students.

The procedure in this lab is written for use with the original CBL system. If you are using CBL 2 or LabPro, the CHEMBIO program can still be used. Updated versions of this program can be downloaded from **www.vernier.com.** For additional information on how to integrate the CBL system into your laboratory, see the Program Introduction.

Some sensors may require the use of an adapter. Students will need to connect the adapter to the sensor before connecting it to the CBL.

As an option to disposing of burned materials in a common container, students might ask about adding water to the materials in the crucibles. Explain that water in the crucibles will affect the mass of the next trial unless the crucible is heated to dryness between trials.

Environmental Chemistry Lab

PROBEWARE LAB

Evaluating Fuels

Calorimetry is a technique that uses an instrument called a *calorimeter* to measure the energy as heat released by materials when they burn. In an ideal calorimeter, a sample is burned in a pure oxygen environment inside a container called a reaction chamber. Surrounding the reaction chamber is a water-filled container. When the sample burns, energy flows into the water, raising its temperature. By measuring the temperature change and the mass of the water, the energy released can be calculated.

Fuels contain varying amounts of energy. For fossil fuels, the most important fuels are those that are the most economical, are the cleanest to burn, and produce the largest amount of energy per unit mass. Amount of energy produced per unit mass is known as the *energy content* of a fuel. Technicians at coal-burning power plants sample fuel from suppliers and test it for energy content and pollutants. The coal with the highest energy content, the least moisture, and the fewest pollutants is the most valuable.

In this lab, you will burn different fuels to find out their energy content. You will use temperature readings, the mass of the heated water, and the specific heat of water to determine the amount of energy flowing into the water. The *specific heat* of a substance is the amount of energy as heat needed to raise 1 g of the substance 1°C. The energy gained by the water, measured in joules, is determined by the equation

$$E = mc_w\Delta T,$$

where E is energy, m is the mass of the water in grams, c_w is the specific heat of water, and ΔT is the temperature change of the water in degrees Celsius. The specific heat of water is a constant equal to 4.184 J/(g°C). Using this data, the amount of energy per gram of fuel burned can be determined.

OBJECTIVES

Use a temperature probe to measure changes in water temperatures generated by burning fuels.

Evaluate the fuel samples to determine which one releases the greatest amount of energy per unit mass.

MATERIALS
• fuels (1–4)

EQUIPMENT
• balance
• CBL system
• clamps, utility (2)
• crucibles (1–4)
• flame lighter
• flasks, 250 mL (1–4)

• link cable
• oven mitt or tongs
• ring stand
• temperature probe
• TI graphing calculator

Evaluating Fuels *continued*

SAFETY ◈ ◈ ◈ ◈ ◈

- Wear safety goggles, gloves, and an apron at all times.

- Be very cautious with the fuels and the flame lighter. Before igniting the fuels, be sure all other combustible materials are cleared away from the setup.

- Glassware is fragile. Notify the teacher of broken glass or cuts. Do not clean up broken glass or spills with broken glass unless the teacher tells you to do so.

Procedure

SETTING UP THE CBL SYSTEM

1. Connect the CBL unit to the graphing calculator, using the link cable. Press the ends of the link cable firmly into the CBL unit and the calculator. Connect the temperature probe to the Channel 1 input.

2. Turn on the graphing calculator and CBL unit. Start the CHEMBIO program. Go the the MAIN MENU.

3. Select SETUP PROBES. Enter "1" as the number of probes. From the SELECT PROBE menu, select TEMPERATURE. Enter "1" as the channel number.

ASSEMBLING THE CALORIMETRY APPARATUS

4. Connect a utility clamp to the ring stand. Place the crucible on the base of the ring stand.

5. Find the mass of the empty flask. Record the mass in **Table 1**. Add approximately 100 mL of water to the flask. Find the mass of the flask and the water. Record this mass in **Table 1**.

6. Secure the water-filled flask to the clamp. Adjust the clamp so that the base of the flask is approximately 2–3 cm higher than the top of the crucible, as shown in **Figure 1**. Use another utility clamp to suspend the temperature probe in the water. The probe should not touch the sides or bottom of the flask.

FIGURE 1 POSITION OF THE CRUCIBLE, FLASK, AND PROBE

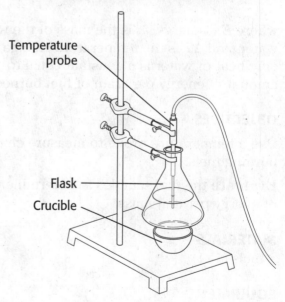

Temperature probe

Flask

Crucible

PREPARING FUEL SAMPLES

7. Prepare your solid fuel samples in small piles.

8. Fill the crucible with a fuel sample. Do not overfill the crucible or pack down the fuel. Record the fuel type in **Table 2.**

9. Place the filled crucible on the balance, and measure its mass. Record the value in **Table 2.** Place the crucible on the base of the ring stand.

COLLECTING DATA

10. Select COLLECT DATA from the MAIN MENU. On the DATA COLLECTION MENU, select MONITOR INPUT. The temperature in degrees Celsius will be displayed on the calculator. Record this initial water temperature in **Table 1.** While you are changing or preparing fuel samples for testing, if the graphing calculator goes into sleep mode, press the ON button to turn it on.

11. Use the flame lighter to ignite the fuel sample. **CAUTION: When using an open flame, tie back long hair and keep loose clothing away from the flame.** Allow the fuel to burn for 3–5 minutes while it heats the water in the flask.

12. Use tongs or an oven mitt to remove the flask from the heat source. Swish the flask to mix the water. Take a final temperature reading, and record it in **Table 1.** When you are finished with data collection, press "+" to end.

13. Find the mass of the crucible containing the burned fuel sample. Record this final crucible and fuel mass in **Table 2.**

14. Dispose of the burned fuel sample according to your teacher's instructions.

15. Repeat steps 8–13 for additional fuel samples as directed by your teacher.

16. Clean up your work area, and wash your hands before leaving the lab.

TABLE 1 WATER MASS AND TEMPERATURE CHANGE

Trial	Empty flask mass (g)	Flask + water mass (g)	Water mass (g)	Initial water temperature (°C)	Final water temperature (°C)	Water temperature change (°C)
1	92.2	197.9	105.7	23.7	38.7	15.0
2	92.2	190.7	98.5	26.3	37.4	11.1
3	92.2	193.0	100.8	25.2	37.8	12.6
4	92.2	194.2	102.0	27.2	55.5	28.3

Evaluating Fuels *continued*

TABLE 2 MASS CHANGE FOR BURNED FUEL

Trial	Fuel type	Initial crucible + fuel mass (g)	Final crucible + fuel mass (g)	Fuel mass change (g)
1	Spruce twigs	30.1	29.1	1.0
2	Craft sticks	31.7	29.9	1.8
3	Paper	27.8	25.4	2.4
4	Alcohol in burner	188.9	187.6	1.3

TABLE 3 ENERGY CALCULATIONS

Trial	m, Water mass (g)	c_w, Specific heat of water (J/g•°C)	ΔT, Water temperature change (°C)	$mc_w\Delta T$, Energy absorbed by water (J)	Fuel mass change (g)	Amount of energy in fuel (J/g)
1	105.7	4.184	15.0	6,634	1.0	6,634
2	98.5	4.184	11.1	4,575	1.8	2,541
3	100.8	4.184	12.6	5,314	2.4	2,214
4	102.0	4.184	28.3	12,078	1.3	9,290

Entries will vary. Sample data are entered in Tables 1–3.

Analysis

1. Organizing Data Calculate the fuel mass change in **Table 2** by substracting the final crucible and fuel mass from the initial crucible and fuel mass. Enter the values in Table 2 and in column 6 of **Table 3.**

2. Organizing Data Calculate the water mass in **Table 1** by subtracting the empty flask mass from the flask and water mass. Enter the values in **Table 1.**

3. Organizing Data Calculate the temperature change of the water in **Table 1** by subtracting the initial water temperature from the final water temperature. Record the results in the last column of **Table 1** and column 4 of **Table 3.**

4. Organizing Data Calculate the energy absorbed by the water for each fuel tested, using the equation $E = mc_w\Delta T$. Use the data columns 2, 3, and 4 of **Table 3,** and record your results in column 5.

5. Analyzing Results Using the data in **Table 3,** find the fuel energy content by dividing the energy absorbed by water by the fuel mass change. Record these values in the last column in **Table 3.**

6. Analyzing Results Rank the fuels tested from least to greatest amount of energy absorbed by the water.

Answers will vary, depending on fuels tested. Sample results are craft sticks,

paper, spruce twigs, and alcohol.

7. Analyzing Results Rank the fuels tested from least to greatest fuel energy content.

Answers will vary, depending on fuels tested. Sample results are paper, craft

sticks, spruce twigs, and alcohol.

Conclusions

1. **Evaluating Methods** Which of the fuel rankings—the one from Analysis question 6 or 7—is more useful? Explain your answer.

   Answers will vary. Most students will find the ranking in Analysis question 7

   more useful. It is based on energy per gram, not the energy produced by

   samples of varying masses.

2. **Evaluating Models** In an ideal calorimeter, all of the released energy is captured by the water. In the calorimeter you used, where did energy flow in addition to the water?

   Energy was transferred to the air, flask, ring support, and crucible.

3. **Drawing Conclusions** Based on energy content, which of the fuels tested do you think is the best?

   Answers will vary. According to the sample results, the alcohol in the burner

   is the best.

4. **Applying Conclusions** What other information, besides energy content, should be considered when determining what makes the best fuel? Given these considerations, which of the fuels tested do you think is the best? Explain.

   Other information might include the amount of pollutants produced, the

   amount of fuel available, and the technology available to obtain the fuel.

   Some high-energy-producing fuels, such as coal, may produce smoke while

   burning, leading students to choose another fuel as being the best.

Extension

Research and Communication Investigate a bomb calorimeter or other type of calorimeter, and prepare a presentation that describes how it works. Include a diagram or model in your presentation.

Name _____ Class _____ Date _____

Topic Introduction

An Introduction to Biological Chemistry

Chemistry is the study of the composition, properties, and structure of matter and the changes that it undergoes. The kinds of matter that chemists study is not limited to things such as metals, gases, and solutions in beakers. Chemists also study the matter that makes up living things. Life processes are chemical and physical changes such as those you have studied in other science courses. These processes include digestion and cell reproduction. **Biological chemistry,** *or biochemistry, is the study of the chemistry of biological processes.* Scientists who work in this field are called *biochemists.*

Knowing how drug molecules interact with biological molecules allows chemists to design new and potentially more effective drugs. Still, biochemists have much to do in order to understand and prevent diseases such as cancer and AIDS, as well as genetic diseases such as sickle-cell anemia and cystic fibrosis. Many biological processes involve large, complex molecules such as DNA. *Genes,* which determine an organism's traits, are made up of DNA. As of 2004, researchers have discovered that the human body contains about 20,000 to 25,000 genes. Understanding how genes determine traits allows chemists to modify traits of organisms. You will learn about some of these modifications in the next section.

A Current Hot Topic: Genetic Engineering

During the twentieth century, research by biochemists has shown how the genetic code determines the traits of an organism. This knowledge has created a new branch of science called genetic engineering. **Genetic engineering** *is the modification of the DNA of an organism in order to change the traits of an organism.* The results can include improving existing traits, eliminating undesirable traits, or adding new traits, often by transferring a gene from a different organism.

EARLY GENETIC MODIFICATION

People have carried out simple forms of genetic engineering for thousands of years. For example, both plants and animals have been crossbred to bring out useful traits, such as greater yields of grain or milk, over the course of generations. Corn, for example, originated as a tall grass with small seed heads that looked like wheat. Thousands of years of crossing corn plants produced the corn with large "ears" that we know today. **Figure 1** on the next page compares the modern cultivated maize plant and the wild grass, *teosinte*, which maize is believed to be derived from. Likewise, modern dogs are descended from wolves. Thousands of years of selective breeding has produced a huge variety of dog breeds, many of which hardly resemble wolves at all!

An Introduction to Biological Chemistry *continued*

FIGURE 1: PRIMITIVE WILD CORN AND MODERN CULTIVATED CORN

Maize (corn), shown on the right, is believed to be derived from the wild grass *teosinte*, shown at left. The heavy seed cob of cultivated corn is much larger than the cob of teosinte due to many generations of selective cultivation.

Primitive corn

Modern corn

wild teosinte

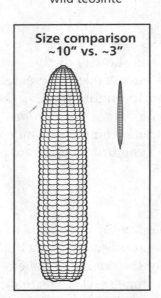

**Size comparison
~10" vs. ~3"**

cultivated
variety of
maize

Maize, right, is believed to be derived from the wild grass teosinte (left), its heavy seed cob being equivalent to teosinte's miniscule ear (left box)

MODERN GENETIC MODIFICATION

In modern genetic engineering, scientists may insert human genes into bacteria to make them synthesize important hormones such as insulin and human growth hormone. These hormones can then be harvested from the bacteria and given to patients who need them. Other genes can make bacteria produce useful drugs that are difficult to synthesize in the laboratory. Plants and fungi can also be genetically engineered to produce hormones and drugs. **Table 1** on the next page lists several examples of modern genetic engineering. Some of these, such as the production of hormones by bacteria and herbicide-resistant plants, are already used commercially. Other examples of genetic engineering are still being developed.

An Introduction to Biological Chemistry *continued*

TABLE 1: SOME APPLICATIONS OF GENETIC ENGINEERING

Organism	Genetic modification	Result
Bacteria, fungi	Human genes are added to bacteria and fungi to produce hormones or antibodies.	Bacteria and fungi synthesize hormones or antibodies.
Food plants	Genes from bacteria that are toxic to insects are added to plants.	Plants synthesize their own pesticides.
Food plants	Genes from cold water fish are added to plants.	Plants become resistant to freezing.
Bacteria, plants	Genes from other organisms that can break down environmental pollutants are added to bacteria or plants.	Bacteria or plants can break down oil spills and chemical pollutants.
Cows, sheep, goats	Genes for the production of hormones or drugs are inserted into the animal embryo.	Animals give milk that contains hormones or drugs.
Humans	Genes for the production of anticancer substances are added to cancerous cells.	Engineered cells multiply in the tumor, helping the body combat the cancer.

GENETICALLY MODIFIED FOODS

Table 1 shows several examples of genetic engineering. Earth's population is about 6.5 billion today and may increase to 8 billion by the year 2020. One of society's biggest challenges is finding ways to feed this huge number of people. The goals of most genetic engineering of food plants—especially corn, soybeans, wheat, and rice—is to increase both crop yields and the nutritional content of foods. Today, about 75% of the soybeans and 35% of the corn in the United States come from genetically engineered plants.

In one example of genetic modification, plants are given genes that make them resistant to herbicides, which are chemicals that kill plants. Then a field of these treated plants can be sprayed with a herbicide that kills all the competing plants (weeds), leaving only the desired crop. In another example, genes from certain bacteria can be inserted into plants, which then synthesize compounds that are toxic to insects: in effect, the plant makes its own pesticide.

Name _____ Class _____ Date _____

An Introduction to Biological Chemistry *continued*

FIGURE 2: COMPARISON OF NURITIONAL CONTENT BETWEEN GENETICALLY ENHANCED CORN AND ORDINARY CORN

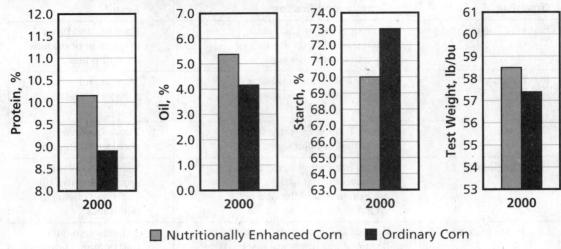

■ Nutritionally Enhanced Corn ■ Ordinary Corn

Genetic engineering can improve the nutritional content of grains such as wheat, corn, and rice that are high in starch but low in protein. Most naturally occurring plant proteins do not contain all of the amino acids that humans need. Genes for the synthesis of additional protein containing amino acids missing from the plant can be added to these plants. The graph in **Figure 2** above compares the protein, oil, and starch content of ordinary corn with corn that has been nutritionally enhanced by genetic engineering. Note that the amounts of protein and oil (fat) in the engineered corn are increased and the starch content is decreased compared with the ordinary corn. Also, the density of the corn in pounds per bushel (lb/bu) is increased, which means that the engineered corn provides more nutrition per unit volume.

CONTROVERSIES OVER GENETIC ENGINEERING

However, there are drawbacks as well as benefits to genetically modified food crops. For example, insect pests could develop resistance to a toxin produced by a plant, and then that toxin would be ineffective against those pests in the future. Some people have raised the concern that added genes could be transferred to nongenetically modified crops. For example, the genes that make crops resistant to herbicides might be passed on to weeds.

Some people can have severe allergic reactions to some foods, such as peanuts. If genes from peanuts were used to genetically modify another plant, people might become allergic to that plant as well. Others fear the possibility that genetically engineered plants, bacteria, and fungi could displace the normal organisms in the ecosystem.

Name _____ Class _____ Date _____

An Introduction to Biological Chemistry *continued*

GENETIC ENGINEERING IN HUMANS

Genetic engineering could be used to cure human genetic diseases such as sickle-cell anemia, cystic fibrosis, and hemophilia. However, the use of genetic engineering to cure or prevent human diseases poses unique difficulties and challenges. For a disease not to develop in a person, genes that prevent the disease must be inserted into a very early embryo or a fertilized egg. As of now, no practical way exists to tell if a genetic disorder exists in early embryos. Many genetic disorders are discovered only after birth. Much research still needs to be done before genetic engineering can be effectively used to cure human genetic diseases. In addition, the potential to screen for some diseases prior to birth also raises a number of ethical questions.

Careers in Biological Chemistry

Genetic engineering is just one of the fast-growing fields within biological chemistry today. The possibilities for careers in this area are almost unlimited.

BIOCHEMIST

Biochemistry today is a vast, wide-open field. A degree in biochemistry can prepare one for an almost unlimited number of careers. In the laboratory, biochemists do basic research in a variety of areas, including genetic engineering, toxicology, medicine, and biotechnology. An advanced degree in biochemistry can also prepare one to teach and do research at a college or university.

PHARMACOLOGIST

Pharmacology is the study of how medicines interact with the human body. Some pharmacologists study the biochemistry of human life processes. Other pharmacologists work for pharmaceutical companies and attempt to design drug molecules that will interact in specific ways with the molecules of the body. In a hospital, a clinical pharmacologist is an expert on the action of drugs and combinations of drugs on patients. Pharmacologists in hospitals are often physicians as well. So, they work with patients by monitoring drug action and the side effects of drugs on patients.

PLANT GENETICIST

A plant geneticist can work to develop new plant varieties by altering the genetic makeup of plants: everything from flowers of different colors to tomatoes that taste better to wheat that is short and wind resistant. A plant geneticist may work in a research laboratory or outdoors, such as on an experimental farm. Some plant geneticists use the techniques of genetic engineering to isolate genes and transfer genes from one species to another. This field usually requires an advanced college degree.

Topic Questions

1. Describe briefly how modern genetic engineering is used to modify an organism.

Sample answer: An organism is genetically engineered by modifying its

genetic code (DNA). Genes from other organisms may be added in order to

produce desired characteristics in the engineered organism.

2. Give two examples of the use of genetic engineering to aid human health.

Student answers may include the synthesis of drugs, vaccines, or human

hormones; or the enhancement of the nutritional value of foods.

3. What is the purpose of creating plants that are resistant to herbicides?

These plants can be sprayed with herbicides that kill all competing plants

but leave the resistant plants unharmed.

4. What problems may arise from planting genetically engineered crops that produce compounds that are toxic to insects?

The toxic compounds may kill beneficial insects as well. Also, the genes may

pass to other plants so that they also become toxic.

5. Use the graphs in **Figure 2** to estimate the increase in percentage of protein and oil and the decrease in percentage of starch in the nutritionally enhanced corn.

The protein content increases by 1.4%, from 8.9% to 10.3%. The oil content

increases by 1.2%, from 4.2% to 5.4%. The starch decreases by 3.0%, from

73.1% to 70.1%.

6. Look at the difference in density between ordinary and nutritionally enhanced corn shown in **Figure 2.** What would be the difference in weight between 1,000 bushels of genetically enhanced corn and ordinary corn?

1,000 bushels of genetically enhanced corn would weigh 1,000 bu $\times$ 58.6

lb/bu = 58,600 lb, and 1,000 bushels of ordinary corn would weigh 1,000 bu $\times$

57.4 lb/bu = 57,400 lb. The difference between them is 58,600 lb – 57,400 lb =

1,200 lb.

Identifying Organic Compounds in Foods

Teacher Notes

TIME REQUIRED one 50-minute class period

LAB RATINGS

Easy ◄—— 1   2   3   4 ——► Hard

Teacher Preparation–2
Student Setup–3
Concept Level–2
Cleanup–3

SKILLS ACQUIRED

Experimenting
Observing
Measuring

SCIENTIFIC METHODS

Make Observations Students will make observations during the experiment.

Analyze the Results Analysis questions 1 and 2 ask students to analyze the results.

Draw Conclusions Conclusions question 1 asks students to draw conclusions.

MATERIALS (PER LAB GROUP)

- 1 L beaker
- hot plate
- droppers (9)
- 10 mL graduated cylinder
- marker
- glass stirring rods (9)
- test tubes (9)
- test-tube rack
- tongs or test-tube holder
- albumin solution (see below)
- Benedict's solution (see below)
- copper sulfate solution
- glucose solution (see below)
- labeling tape
- sodium hydroxide solution
- Sudan III solution (see below)
- unknown solution (see below)
- vegetable oil
- distilled water

Albumin, Benedict's, and Sudan III solutions are available from WARD'S and most supply companies. To make the unknown solution, add 5 g of cottage cheese to 100 mL of distilled water, and stir until mixed thoroughly. Purchase glucose in powder form, and prepare a 10% solution (10 g in a final volume of 100 mL of water).

Identifying Organic Compounds in Foods *continued*

Purchase albumin in powder form, and prepare a 10% solution (10 g in a final volume of 100 mL of water).

Prepare an 8% sodium hydroxide solution (8 g in a final volume of 100 mL of water) and a 1% copper(II) sulfate solution (1 g in a final volume of 100 mL of water) separately. Alternatively, purchase biuret reagent from a biological supply house such as WARD'S.

To prepare a 2% stock solution of Sudan III, add 2 g of Sudan III and dilute to a final volume of 100 mL with undiluted ethyl alcohol. Just before using, mix the stock solution with an equal amount of 45% ethyl alcohol.

SAFETY CAUTIONS

Students should wear a lab apron, protective gloves, and safety goggles at all times during this investigation. Instruct students to handle the solutions used in this lab carefully. The solutions can injure skin and eyes or stain skin and clothing. Do not use Sudan III solution in the same room with an open flame.

If any chemical gets in a student's eye, flush the eye with water for at least 15 minutes and seek immediate medical attention. Students should use tongs to handle hot test tubes and should not handle the hot plate with wet hands or while near water.

DISPOSAL

Neutralize the sodium hydroxide solution with baking soda or vinegar, and pour down the drain. Dispose of albumin and dye solutions in aqueous chemical waste containers.

TIPS AND TRICKS

To save time, you may want to have one-third of the students perform test 1 while the other students perform tests 2 and 3.

Biological Chemistry Lab

Identifying Organic Compounds in Foods

Carbohydrates, proteins, and lipids are nutrients that are essential to all living things. Some foods, such as table sugar, contain only one of these nutrients. Most foods, however, contain mixtures of proteins, carbohydrates, and lipids. You can confirm this by reading the information in the "Nutrition Facts" box found on any food label.

In this investigation, you will use chemical substances, called indicators, to identify the presence of specific nutrients in an unknown solution. By comparing the color change an indicator produces in the unknown food sample with the change it produces in a sample of known composition, you can determine whether specific organic compounds are present in the unknown sample.

Benedict's solution is used to determine the presence of monosaccharides, such as glucose. A mixture of sodium hydroxide and copper sulfate determine the presence of some proteins (this procedure is called the biuret test). Sudan III is used to determine the presence of lipids.

OBJECTIVES

Determine whether specific nutrients are present in a solution of unknown composition.

Perform chemical tests using substances called *indicators*.

MATERIALS

- albumin solution
- Benedict's solution
- copper sulfate solution
- glucose solution
- labeling tape
- sodium hydroxide solution
- Sudan III solution
- unknown solution
- vegetable oil
- distilled water

EQUIPMENT

- 1 L beaker
- hot plate
- droppers (9)
- 10 mL graduated cylinder
- marker
- glass stirring rods (9)
- test tubes (9)
- test-tube rack
- tongs or test-tube holder

SAFETY ⬦ ⬦ ⬦ ⬦ ⬦ ⬦ ⬦ ⬦ ⬦

- Put on a lab apron, safety goggles, and gloves.

- In this lab, you will be working with chemicals that can harm your skin and eyes or stain your skin and clothing. If you get a chemical on your skin or clothing, wash it off at the sink while calling to your teacher. If you get a chemical in your eyes, immediately flush it out at the eyewash station while calling to your teacher.

- Do not touch the hot plate. Do not touch the test tubes with your hands.

- Do not plug or unplug the hot plate with wet hands.

- Do not use Sudan III solution in the same room with an open flame.

Procedure

As you perform each test, record your data in your lab report, organized in a table like the one on the next page.

TEST 1

1. Make a water bath by filling a 1 L beaker half full with water. Then put the beaker on a hot plate and bring the water to a boil.

2. While you wait for the water to boil, label one test tube "1-glucose," label the second test tube "1-unknown," and label the third test tube "1-water." Using the graduated cylinder, measure 5 mL of Benedict's solution and add it to the "1-glucose" test tube. Repeat the procedure, adding 5 mL of Benedict's solution each to the "1-unknown" test tube and "1-water" test tube.

3. Using a dropper, add 10 drops of glucose solution to the "1-glucose" test tube. Using a second dropper, add 10 drops of the unknown solution to the "1-unknown" test tube. Using a third dropper, add 10 drops of distilled water to the "1-water" test tube. Mix the contents of each test tube with a clean stirring rod. (It is important not to contaminate test solutions by using the same dropper or stirring rod in more than one solution. Use a different dropper and stirring rod for each of the test solutions.)

4. When the water boils, use tongs to place the test tubes in the water bath. Boil the test tubes for 1 to 2 minutes.

5. Use tongs to remove the test tubes from the water bath and place them in the test-tube rack. As the test tubes cool, an orange or red precipitate will form if large amounts of glucose are present. If small amounts of glucose are present, a yellow or green precipitate will form. Record your results in your data table.

SAMPLE DATA TABLE: IDENTIFICATION OF SPECIFIC NUTRIENTS BY CHEMICAL INDICATORS

Test	Nutrient in test solution	Nutrient category (protein, lipid, etc.)	Result for known sample	Result for unknown sample	Result for distilled water
1	glucose	carbohydrate	red or orange precipitate	no change	no change
2	albumin	protein	pink-purple color	pink-purple color	no change
3	vegetable oil	lipid	pink color	no change	no change

TEST 2

6. Label one clean test tube "2-albumin," label a second test tube "2-unknown," and label a third test tube "2-water." Using a dropper, add 40 drops of albumin solution to the "2-albumin" test tube. Using a second dropper, add 40 drops of unknown solution to the "2-unknown" test tube. Using a third dropper, add 40 drops of water to the "2-water" test tube.

7. Add 40 drops of sodium hydroxide solution to each of the three test tubes. Mix the contents of each test tube with a clean stirring rod.

8. Add a few drops of copper sulfate solution, one drop at a time, to the "2-albumin" test tube. Stir the solution with a clean stirring rod after each drop. Note the number of drops required to cause the color of the solution in the test tube to change. Then add the same number of drops of copper sulfate solution to the "2-unknown" and "2-water" test tubes.

9. Record your results in your data table.

TEST 3

10. Label one clean test tube "3-vegetable oil," label a second test tube "3-unknown," and label a third test tube "3-water." Using a dropper, add 5 drops of vegetable oil to the "3-vegetable oil" test tube. Using a second dropper, add 5 drops of the unknown solution to the "3-unknown" test tube. Using a third dropper, add 5 drops of water to the "3-water" test tube.

11. Using a clean dropper, add 3 drops of Sudan III solution to each test tube. Mix the contents of each test tube with a clean stirring rod.

12. Record your results in your data table.

13. Clean up your materials and wash your hands before leaving the lab.

Analysis

1. Analyzing Methods What are the experimental controls in this investigation?

The controls are three known foods: glucose, albumin, and vegetable oil.

2. Analyzing Methods Explain how you were able to use the color changes of different indicators to determine the presence of specific nutrients in the unknown substance.

The presence of different nutrients in foods can be detected by color

changes in chemical indicators specific for that nutrient.

3. Analyzing Methods List four potential sources of error in this investigation.

Possible sources of error include inaccurate measurements, impurities in the

solutions and samples, unwashed equipment, and, in test 1, inadequately

heated or cooled samples.

Conclusions

1. Drawing Conclusions Based on the results you recorded in your data table, identify the nutrient or nutrients in the unknown solution.

Students' results should reveal that the main nutrient in the unknown is

protein, as evidenced by the purple color in the protein test, test 3. Nonfat

cottage cheese also contains a small amount of sugars.

Measuring the Release of Energy from Sucrose

Teacher Notes

TIME REQUIRED two days

LAB RATINGS   Easy ◄— 1   2   3   4 —► Hard

   Teacher Preparation–3
   Student Setup–3
   Concept Level–2
   Cleanup–2

SKILLS ACQUIRED

   Experimenting
   Measuring
   Predicting
   Organizing and Analyzing Data

SCIENTIFIC METHODS

Make Observations Students will make observations during the course of the experiment.

Analyze the Results Analysis question 1 requires students to analyze their results.

Draw Conclusions Conclusions questions 1–4 ask students to draw conclusions from their data.

MATERIALS (PER LAB GROUP)

- lukewarm water, 800 mL
- limewater, 100 mL
- sucrose, 150 g
- dried yeast package
- insulated bottles, 500 mL (2)
- beaker, 1000 mL
- flask, 250 mL
- pieces of rubber tubing, 40 cm (2)
- glass stirring rod
- wax pencil
- two-hole stoppers with a thermometer inserted in one hole and a 10 cm piece of glass inserted in the other hole (2)

To insert glass tubing into the stopper, first be sure that the ends of the tubing are fire-polished and that the outside diameter of the tubing is only slightly larger than the diameter of the hole in the stopper. Lubricate the holes in the stopper with glycerin or water. Then, wearing thick leather gloves or with both hands protected by multiple layers of cloth, gently insert the tubing into a hole without twisting or turning the tube or stopper. Do not force the glass tubing; it can shatter. Repeat for the thermometer.

| Measuring the Release of Energy from Sucrose *continued*

SAFETY CAUTIONS

Make sure students wear safety goggles and a lab apron. Caution students to exercise care with all glassware.

DISPOSAL

Solutions should be disposed of in aqueous chemical waste containers.

NOTES ON TECHNIQUE

Students will need to return to the lab at various times during the school day to take additional data readings.

If enough oxygen is dissolved in the solution and the temperature is high enough, cellular respiration may occur in addition to fermentation. The occurrence of cellular respiration will be marked by a steeper rise in temperature in the "Experimental" bottle.

Microcomputer-Based Laboratory (MBL) probes and Calculator-Based Laboratory (CBL) probes can be used to measure changes in temperature. Have students insert a thermistor (temperature probe) into one of the holes of the two-hole stopper. Provide modeling clay to seal the hole so that air cannot enter and so that heat will remain inside.

SAMPLE DATA

Student graphs should resemble the following:

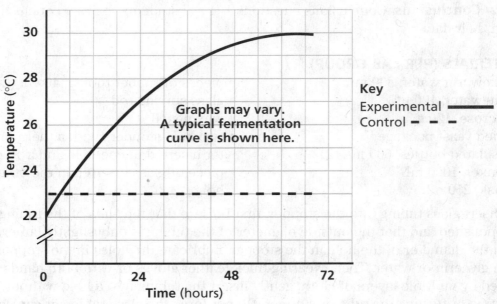

Change in Temperature in Bottles

Biological Chemistry Lab

Measuring the Release of Energy from Sucrose

The carbohydrate *sucrose*, or table sugar, is made by plants to store energy. All living things must have energy to power their cells. That energy is obtained from organic molecules such as sucrose through the processes of *fermentation* and *cellular respiration*. Fermentation, which occurs in the absence of oxygen, releases a relatively small amount of energy. Cellular respiration, which occurs in the presence of oxygen, releases a great deal of energy. Because energy conversions are not 100 percent efficient, some of the energy released is in the form of heat. The addition of energy as heat to a system results in an increase in temperature.

Another product of both fermentation and cellular respiration is carbon dioxide. A solution called *limewater* can be used to indicate the presence of carbon dioxide. Limewater, which is normally clear, turns cloudy in the presence of carbon dioxide.

OBJECTIVES

Predict how the temperature of a solution will change as a chemical process occurs in the solution.

Relate a change in temperature to the release of energy from sucrose.

MATERIALS
- lukewarm water, 800 mL
- limewater, 100 mL
- sucrose, 150 g
- dried yeast package

EQUIPMENT
- safety goggles
- lab apron
- insulated bottles, 500 mL (2)
- beaker, 1000 mL
- flask, 250 mL
- pieces of rubber tubing, 40 cm (2)
- glass stirring rod
- wax pencil
- two-hole stoppers with a thermometer inserted in one hole and a 10 cm piece of glass inserted in the other hole (2)

SAFETY ◆ ◆ ◆

- Always wear safety goggles and a lab apron to protect your eyes and clothing.
- Glassware is fragile. Notify the teacher of broken glass or cuts. Do not clean up broken glass or spills with broken glass unless the teacher tells you to do so.

Procedure

1. Prepare a data table in which to record time and temperature values for the experimental bottle and the control bottle.

2. Put on safety goggles and a lab apron. Using a wax pencil, label two insulated bottles with the initials of your group's members. Label one of the bottles "Control" and the other "Experimental."

3. In a 1000 mL beaker, dissolve 150 g of sucrose in 800 mL of lukewarm water. Pour half of the sucrose solution into the insulated bottle marked "Control."

4. Add one package of fresh yeast to the sucrose solution remaining in the beaker, and stir. Pour the sucrose-yeast mixture into the bottle marked "Experimental."

5. Set up the two insulated bottles, two 250 mL flasks with limewater, and two 40 cm pieces of rubber tubing as shown in **Figure 1** below. Adjust the thermometers until they extend down into the solutions in the bottles and the liquid column can be seen above the stopper. *Note: Be sure that the glass tubing does not touch the solutions in the bottles.*

FIGURE 1

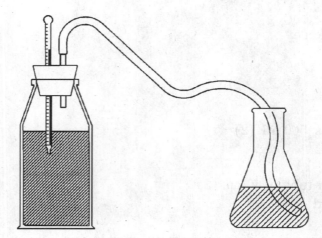

6. Record the time and the initial temperature of the solution in each bottle in your data table.

What do you predict will happen in each bottle over the next 48 hours?

Answers may vary. Students should predict that the temperature in the

"Control" bottle will not change while the temperature in the

"Experimental" bottle will rise as the yeast breaks down sucrose and

releases energy stored in the sucrose.

7. Clean up your work area and wash your hands before leaving the lab.

8. Continue to record the temperature in each bottle for 48 hours. Take turns with your lab partner during the school day to take temperature data between class periods. Record as many readings as possible during the next 48 hours. What happened in each flask of limewater?

The limewater in the flask paired with the "Experimental" bottle turned

cloudy. The limewater in the flask paired with the "Control" bottle stayed

clear.

9. Dispose of your materials according to the directions from your teacher.

10. Make a line graph of the data. Plot time on the x-axis (horizontal axis) and temperature on the y-axis (vertical axis). Use two different colors to graph the results from each bottle. Plot a point on the graph that corresponds to the temperature at each time. The curve of the experimental data points will show the yeast's energy production. Complete the graph by drawing curves through the plotted points.

Analysis

1. Analyzing Results What are some indications that a chemical reaction was taking place inside one of the insulated bottles?

Bubbles appeared in the limewater, which turned cloudy, and there was a rise

in temperature.

2. Analyzing Methods What is the purpose of the glass tube and rubber tubing in this experiment?

The glass tube enables any gases produced to be released from the bottles

and prevents the rubber stopper from popping off due to pressure buildup.

The rubber tubing attached to the glass tube enables the gas to bubble

through limewater, showing that the gas is carbon dioxide.

Name _____ Class _____ Date _____

Measuring the Release of Energy from Sucrose *continued*

Conclusions

1. Drawing Conclusions How would you explain the change in temperature observed during the experiment?

The temperature increased in the "Experimental" bottle because energy was

released as the yeast broke down sucrose. The temperature varied very little

in the "Control" bottle because there was no yeast to break down sucrose.

2. Drawing Conclusions How could you alter this experiment to show that the energy is released from sucrose?

Answers may vary. Students should suggest adding a third setup to the

experiment, in which there is yeast but no sucrose.

3. Drawing Conclusions How would you explain what happens to the lime-water in the two setups?

The limewater turns cloudy in the "Experimental" setup because carbon

dioxide gas is released by the yeast during fermentation. The limewater

remains clear in the "Control" setup because there is no yeast in the bottle

and no carbon dioxide is produced.

4. Inferring Conclusions Which process, fermentation or cellular respiration, caused the changes observed in this experiment? How could you test your hypothesis?

Answers may vary. Students should realize that oxygen cannot enter the

systems in the bottles and that any oxygen present at the outset would be

depleted rapidly. Therefore, fermentation is the most likely cause of the

changes observed during the experiment. This hypothesis could be tested by

testing the solution in the "Experimental" bottle for the presence of alcohol,

which is produced by fermentation but not by cellular respiration.

Extension

1. **Further Inquiry** Soon after the yeast is placed in the sucrose solution, take one drop of the solution and place it on a slide. Observe the yeast with a microscope under high power, and estimate the number of cells seen. Repeat the above process on the second and third days. Relate your findings to the energy-releasing process.

2. **Research and Writing** Look up the meanings of the terms *exothermic*, *endothermic*, *exergonic*, and *endergonic*. Write a paragraph that describes the processes of fermentation and cellular respiration using the appropriate terms from the list above.

Diffusion and Cell Membranes

Teacher Notes

TIME REQUIRED two to three 50-minute class periods

LAB RATINGS

Easy ←———— 1   2   3   4 ————→ Hard

Teacher Preparation–2
Student Setup–3
Concept Level–4
Cleanup–3

SKILLS ACQUIRED

Modeling
Observing
Predicting
Measuring

SCIENTIFIC METHODS

Make Observations Students will make observations during the course of the experiment.

Form a Hypothesis Procedure step 8 asks students to form a hypothesis.

Analyze the Results Analysis questions 1–4 require students to analyze their results.

Draw Conclusions Conclusions questions 1 and 2 asks students to draw conclusions from their data.

MATERIALS (PER LAB GROUP)

- corn syrup, 1 bottle
- distilled water
- raw eggs (2)
- paper towels (2)
- vinegar (400 mL)
- thermometer
- 250 mL beakers (6)
- 600 mL beakers (2)
- glass stirring rod
- tablespoon or tongs

SAFETY CAUTIONS

Make sure students wear safety goggles and a lab apron. To avoid possible *Salmonella* poisoning from the raw eggs, be sure that all students avoid placing their hands near their mouths. Also ensure that they thoroughly wash their hands and clean the work surfaces after each day's procedures. Have students carefully wash all glassware after use with the eggs. Use an antibacterial soap if possible.

NOTES ON TECHNIQUE

This inquiry can be done as a demonstration, with students involved in the setup. This would be less costly.

Review the terms *solution* and *solute* with the students. A solution is a homogeneous mixture of two or more substances. The substance present in the greater amount (usually a liquid) is the solvent. The substance present in the lesser amount is the solute. Explain that vinegar is a solution.

Biological Chemistry Lab

Diffusion and Cell Membranes

Some chemicals can pass through the cell membrane, but others cannot. Not all chemicals can pass through a cell membrane with equal ease. The cell membrane determines which chemicals can diffuse into or out of a cell.

As chemicals pass into and out of a cell, they move from areas of high concentration to areas of low concentration. Cells in *hypertonic* solutions have solute concentrations lower than the solution that bathes them. This concentration difference causes water to move out of the cell into the solution. Cells in *hypotonic* solutions have solute concentrations greater than the solution that bathes them. This concentration difference causes water to move from the solution into the cell. The movement of water into and out of a cell through the cell membrane is called *osmosis*.

In this lab, you will use eggs with a dissolved shell as a model for a living cell. You will then predict the results of an experiment that involves the movement of water through a membrane.

OBJECTIVES

Explain changes that occur in a cell as a result of diffusion.

Distinguish between hypertonic and hypotonic solutions.

MATERIALS
- corn syrup, 1 bottle
- distilled water
- raw eggs (2)
- paper towels (2)
- vinegar (400 mL)

EQUIPMENT
- thermometer
- 250 mL beakers (6)
- 600 mL beakers (2)
- glass stirring rod
- tablespoon or tongs

SAFETY

- Always wear safety goggles and a lab apron to protect your eyes and clothing. Do not touch or taste any chemicals. Know the locations of the emergency shower and eyewash station and how to use them.

- If you get a chemical on your skin or clothing, wash it off at the sink while calling to the teacher. Notify the teacher of a spill. Spills should be cleaned up promptly, according to your teacher's directions.

- Glassware is fragile. Notify the teacher of broken glass or cuts. Do not clean up broken glass or spills with broken glass unless the teacher tells you to do so.

Name _____ Class _____ Date _____

| Diffusion and Cell Membranes *continued*

Procedure

1. Label one 600 mL beaker "Egg 1: water" and the other 600 mL beaker "Egg 2: syrup." Also label the beakers with the initials of each member of your group. Measure the mass of each of two eggs to the nearest 0.1 g, and record your measurements in **Table 1** below. **CAUTION: When handling raw eggs, clean up any material from broken eggs immediately. Wash your hands with soap and water after handling the eggs.**

2. Put on safety goggles and a lab apron. Pour 200 mL of vinegar into each labeled beaker. Using a tablespoon or tongs, place an egg into each beaker. *Note: Always return each egg to the same beaker.*

TABLE 1: EGGS IN VINEGAR

Egg	Mass of fresh egg with shell	Observations after 24 h	Mass after 24 h in vinegar
1	x	Shell is dissolved.	$>x$
2	x	Shell is dissolved.	$>x$

3. Place a 250 mL beaker containing 100 mL of water on each egg to keep it submerged as shown in **Figure 1** below. Add more vinegar if the egg is not covered by the vinegar already in the beaker. If some vinegar spills over when the 250 mL beaker is placed on the egg, carry the 600 mL beaker carefully to a sink and pour vinegar some out. Store your beakers for 24 hours in the area specified by your teacher.

4. Clean up your work area and wash your hands before leaving the lab.

FIGURE 1

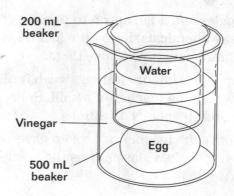

200 mL beaker

Water

Vinegar

Egg

500 mL beaker

5. After 24 hours, observe the eggs. Record your observations in **Table 1.**

6. Put on safety goggles and a lab apron. Label two separate sheets of paper towel "Egg 1" and "Egg 2," respectively. Pour the vinegar from the beakers into a sink. Using a tablespoon or tongs, remove the eggs and rinse them with water. Place each egg on the appropriately labeled paper towel. Measure the mass of each egg, and record the measurement in **Table 1.**

7. Return Egg 1 to its beaker, and add water until the egg is covered. Return Egg 2 to its beaker, and add corn syrup until the egg is covered. Store the beakers and eggs in the same place as before for 24 hours. Clean up your work area, and wash your hands before leaving the lab.

8. Predict how the mass of each egg will change after 24 hours in each liquid. (HINT: An egg is surrounded by a membrane. Inside the membrane, the egg white consists mainly of water and dissolved protein. The yolk consists mainly of fat and water. Syrup is sugar dissolved in water. The protein, fat, and sugar are solutes.) Record your predictions in **Table 2** below.

What will have occurred if your egg gains or loses mass?

Diffusion (or, specifically, osmosis) will have occurred.

9. Observe your eggs after 24 hours. Record your observations in **Table 2.** Measure and record the final masses of the two eggs.

TABLE 2: EGGS SOAKED IN TWO LIQUIDS

Egg	Liquid	Prediction of mass of fresh egg with shell	Observations after 24 h	Mass after 24 h in vinegar
1	water	Predictions may vary.	Egg is very full.	increased
2	syrup	Predictions may vary.	Egg looks smaller. Membrane is looser.	decreased

10. Dispose of your materials according to the directions given by your teacher.

11. Clean up your work area, and wash your hands before leaving the lab.

Analysis

1. Analyzing Results What caused the change in appearance in Egg 1 after it was soaked in the water?

Water moved through the membrane into the egg (or osmosis occurred,

allowing water into the egg).

2. Applying Concepts Why did the mass of the egg increase after soaking in the vinegar solution?

Water moved from the vinegar into the egg.

3. Analyzing Results How did your results in step 9 compare with the prediction you made in step 8?

Answers may vary, depending on the students' predictions.

4. Analyzing Results What material seems to have moved through the membrane of Egg 2 after it was soaked in the syrup? In what direction did it move?

Water moved through the membrane out of the egg.

Conclusions

1. Drawing Conclusions Which egg was in a hypertonic solution? Explain.

Egg 2 was in a hypertonic solution because water diffused from the egg, thus

lowering its mass.

2. Drawing Conclusions Which egg was in a hypotonic solution? Explain.

Egg 1 was in a hypotonic solution because water diffused into the egg, thus

increasing its mass.

Extension

1. Further Inquiry Design an experiment to see what results can be obtained by placing living yeast cells in hypertonic and hypotonic solutions.

2. Further Inquiry Look up the chemical compositions of egg shell and of vinegar. Use this information to write a chemical equation that explains how egg shell dissolves in vinegar.

Observing Enzyme Detergents

Teacher Notes

TIME REQUIRED 45 minutes for Day 1; 30 minutes for Day 2; 15 minutes for Day 3

LAB RATINGS Easy ← 1 2 3 4 → Hard

Teacher Preparation–2
Student Setup–3
Concept Level–4
Cleanup–3

SKILLS ACQUIRED

Asking questions
Collecting data
Experimenting
Identifying patterns
Inferring
Interpreting
Measuring
Organizing and analyzing data

SCIENTIFIC METHODS

Make Observations Students will make observations during the course of the experiment.

Analyze the Results Analysis questions 1, 2, and 4 require students to analyze their results.

Draw Conclusions Conclusions question 1 asks students to draw conclusions from their data.

MATERIALS (PER LAB GROUP)

- 1 g each of 5 brands of laundry detergent
- 18 g regular instant gelatin or 1.8 g sugar-free instant gelatin
- 0.7 g Na_2CO_3
- pH paper
- plastic wrap
- tape
- 50 mL boiling water
- 50 mL distilled water
- balance
- 150 mL beaker
- 50 mL beakers (6)
- glass stirring rod
- graduated cylinder, 10 mL
- metric ruler
- pipet with bulb
- test-tube rack
- test tubes (6)
- thermometer
- oven mitt or tongs
- wax pencil

SAFETY CAUTIONS

Make sure students wear safety goggles and a lab apron. Caution students to avoid burns by working carefully when heating and pouring boiling water.

NOTES ON TECHNIQUE

Before starting the lab, ask students why enzymes are added to detergents (*to help break down proteins and other substances from food that may stain clothing*). Ask students why detergent enzymes are stable during the hot water cycle (*the enzymes found in commercial laundry soap have been genetically engineered*). The rate of gelatin hydrolysis is slower in instant gelatin that contains sugar than in sugar-free gelatin.

Students should bring in labeled samples (3–5 tablespoons) of each of five different laundry detergents. Labels should include active ingredients listed on the container. Students should have a control test tube with 15 drops (1 mL) of water (no detergent) added to the gelatin surface.

An example of a question students may want to explore is: How can you tell if a detergent contains enzymes? Require students to present a written procedure for their experiment and a list of all safety precautions before allowing them to gather materials for the lab.

When students add the detergent solutions to the gelatin, the mixture will foam. During this reaction, carbon dioxide gas is released. The addition of washing soda (Na_2CO_3) to the gelatin raises the pH of the gelatin from 4 to 8, which is the optimum pH for protease activation in the detergent samples. Have students use a wax pencil to mark the test tubes at the uppermost level of the cooled gelatin in each tube. They will use this mark to measure the hydrolysis of the gelatin each day. Label the test tubes 1–6. To prepare a 10% solution of laundry detergent, students should dissolve 1 g of detergent in 9 mL of distilled water. Have students record the pH for each numbered detergent sample. Students can measure protein hydrolysis after 24 hours by using a wax pencil to draw a second line at the top of the gelatin layer and then measuring the distance (in millimeters) between the first line and the second line. This indicates the amount of hydrolysis of the protein in the gelatin by the enzymes in the detergent.

SAMPLE PROCEDURE

1. Prepare a 10% solution of each of five different detergents.

2. Test the pH of each detergent solution with pH paper. Record the pH for each sample in a data table.

3. Add about 15 drops (1 mL) of the first detergent solution to the gelatin surface of the first test tube. Repeat for each of the other samples. Draw a wax pencil line at the top of the gelatin surface of each test tube. Reseal the test tubes and place in a test-tube rack for observation.

4. Add 15 drops of water to the gelatin surface of the sixth test tube.

5. Put the test tubes aside for 24 hours at room temperature.

6. Draw another wax pencil line at the top of the gelatin layer for each test tube. Measure the distance in millimeters between the first line and the second line. Record the data in a data table.

Name _____ Class _____ Date _____

Observing Enzyme Detergents *continued*

Name _____ Class _____ Date _____

Observing Enzyme Detergents

Enzymes are substances that speed up chemical reactions. Each enzyme operates best at a particular pH and temperature. Substances on which enzymes act are called substrates. Many enzymes are named for their substrates. For example, a protease is an enzyme that helps break down proteins. In this lab, you will investigate the effectiveness of laundry detergents that contain enzymes.

OBJECTIVES

Recognize the function of enzymes in laundry detergents.

Relate temperature and pH to the activity of enzymes.

MATERIALS

- 1 g each of 5 brands of laundry detergent
- 18 g regular instant gelatin or 1.8 g sugar-free instant gelatin
- 0.7 g Na_2CO_3
- pH paper
- plastic wrap
- tape
- 50 mL boiling water
- 50 mL distilled water

EQUIPMENT

- balance
- 150 mL beaker
- 50 mL beakers (6)
- glass stirring rod
- 10 mL graduated cylinder
- metric ruler
- pipet with bulb
- test-tube rack
- 6 test tubes
- thermometer
- oven mitt or tongs
- wax pencil

SAFETY

- Always wear safety goggles and a lab apron to protect your eyes and clothing. Do not touch or taste any chemicals. Know the location of the emergency shower and eyewash station and how to use them.

- If you get a chemical on your skin or clothing, wash it off at the sink while calling to the teacher. Notify the teacher of a spill. Spills should be cleaned up promptly, according to your teacher's directions.

- Glassware is fragile. Notify the teacher of broken glass or cuts. Do not clean up broken glass or spills with broken glass unless the teacher tells you to do so.

Pre-Lab Preparation

Based on the objectives for this lab, write a question you would like to explore about enzyme detergents.

Procedure

PART A: MAKE A PROTEIN SUBSTRATE

1. Put on safety goggles and a lab apron.

2. **CAUTION: Use an oven mitt or tongs to handle heated glassware.** Put 18 g of regular (1.8 g of sugar-free) instant gelatin in a 150 mL beaker. Slowly add 50 mL of boiling water to the beaker, and stir the mixture with a stirring rod. Test and record the pH of this solution.

3. Very slowly add 0.7 g of Na_2CO_3 to the hot gelatin while stirring. Note any reaction. Test and record the pH of this solution.

4. Place six test tubes in a test-tube rack. Pour 5 mL of the gelatin-Na_2CO_3 mixture into each tube. Use a pipet to remove any bubbles from the surface of the mixture in each tube. Cover the tubes tightly with plastic wrap and tape. Cool the tubes, and store them at room temperature until you begin Part C. Complete step 12.

PART B: DESIGN AN EXPERIMENT

5. Work with members of your lab group to explore the question you wrote in the Pre-Lab Preparation. To explore the question, design an experiment that uses the materials listed for this lab.

   As you design your experiment, decide the following:

   a. what question you will explore

   b. what hypothesis you will test

   c. what detergent samples you will test

   d. what your control will be

   e. how much of each solution to use for each test

   f. how to determine if protein is breaking down

   g. what data to record in your data table

6. Write a procedure for your experiment. Make a list of all the safety precautions you will take. Have your teacher approve your procedure and safety precautions before you begin the experiment.

PART C: CONDUCT YOUR EXPERIMENT

7. Put on safety goggles and a lab apron.

8. Make a 10% solution of each laundry detergent by dissolving 1 g of detergent in 9 mL of distilled water.

9. Set up your experiment. Repeat step 12.

10. Record your data after 24 hours.

PART D: CLEANUP AND DISPOSAL

11. Dispose of solutions, broken glass, and gelatin in the designated waste containers. Do not pour chemicals down the drain or put lab materials in the trash unless your teacher tells you to do so.

12. Clean up your work area and all lab equipment. Return lab equipment to its proper place. Wash your hands thoroughly before leaving the lab and after finishing all work.

Analysis

1. **Analyzing Methods** Suggest a reason for adding Na_2CO_3 to the gelatin solution.

   The washing soda increases the pH of the gelatin from 4 to 8—the optimum

   pH for enzyme activity in this reaction.

2. **Analyzing Results** Make a bar graph of your data. Plot the amount of gelatin broken down (change in the depth of the gelatin) on the y-axis and detergent on the x-axis.

   Students' graphs may vary.

Conclusions

1. **Drawing Conclusions** What conclusions did your group infer from the results? Explain.

   Answers may vary. For example, enzymes operate best at a certain tempera-

   ture and pH.

Extension

1. Further Inquiry Write a new question about enzyme detergents that could be explored with another investigation.

Answers may vary. For example: Are enzymes in detergent stable in the

presence of bleach?

Do research in the library or media center to answer these questions:

2. What other household products contain enzymes, and what types of enzymes do they contain?

Answers may vary. For example, dairy digestive supplements contain the

enzyme lactase.

3. What type of organic compound is broken down by each enzyme that you identified?

Answers may vary. For example, lactase breaks down lactose, the sugar in

milk.

An Introduction to Food Chemistry

When people eat food, they don't usually think about chemistry. They may begin to think of chemistry when they read a list of ingredients on a food package and see terms such as folic acid, BHT, or monosodium glutamate. People may react by saying that these are "chemicals" in the food. However, as you have learned in previous science classes, food is matter, and like all other matter, it is composed of chemical compounds.

The major food chemicals are proteins, fats, carbohydrates, water, mineral ions, and flavoring agents. **Food chemistry** *is the identification of the components of food and the study of the changes that take place when food is stored, prepared, and cooked.* Food chemists find out how processing and cooking change the taste, appearance, aroma, and nutrient content of food. Some food chemists study how substances used in processing affect the final food product. Another branch of food chemistry is *flavor chemistry*, in which chemists analyze the natural substances that give flavor and aroma to food and the ways that their molecular structures produce certain tastes and smells.

A Current Hot Topic: Flavor, Aroma, and Molecular Structure

Doesn't pizza smell good right out of the oven? Does the anticipation of its taste make your mouth water? How about a freshly baked muffin? It certainly smells and tastes a lot different from the raw batter. There is a lot of chemistry going on when you smell or taste foods.

THE DIFFERENCE BETWEEN "TASTE" AND "FLAVOR"

Food chemists use the words *aroma* and *flavor* instead of "smell" and "taste" because they understand each of these four words in different ways. You may have learned that humans recognize four fundamental tastes—sweet, sour, salty, and bitter. Some scientists would add a fifth fundamental taste called *umami,* a Japanese word for the taste of meat broth. This taste is associated with the presence of the amino acid glutamic acid. Monosodium glutamate, MSG for short, has the umami taste. Other research has pointed to the existence of even a sixth fundamental taste: the taste of polyunsaturated fatty acids, which are fatty acids that have more than one double bond between carbon atoms in the chain.

The term *flavor* involves much more than the fundamental tastes. Flavor refers to the total experience you have when you taste food, which is a combination of taste, aroma, and chemical irritation.

Aroma is our perception of the volatile substances that reach the *olfactory,* or smell, receptors in the nasal cavities. Some scientists estimate that 90% of what we call flavor is really the aroma of food. Many of the volatile compounds that contribute to flavor enter the nasal cavities from the mouth, not through the nostrils, as shown in **Figure 1** on the next page. So, the sensations of aroma and flavor are not easily separated.

Name _____ Class _____ Date _____

An Introduction to Food Chemistry *continued*

FIGURE 1: THE NASOPHARYNGEAL PASSAGE

Food chemicals enter the nasopharyngeal passage, which links the mouth and nose, and are detected by olfactory (smell) receptors. Smells contribute much to the overall "flavor" experience of a food.

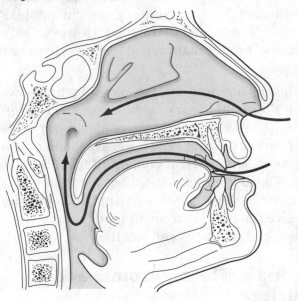

FLAVOR INCLUDES CHEMICAL IRRITATION

In addition to taste and aroma, foods such as black pepper, hot chiles, ginger, and cinnamon have a "hot" flavor because they contain chemically irritating compounds. These compounds irritate the taste and smell receptors, causing a burning sensation. The "heat" of a food depends on which substances are present and how concentrated the substances are. Scientists have suggested that people enjoy hot food because the body releases endorphins in reaction to the "heat." Endorphins produce a sensation of pleasure. However, at some concentrations, these compounds can cause physical pain and can be toxic.

WHY ARE FLAVORS ADDED?

A lot of food chemists work in the food industry developing appealing flavors for processed food. These "flavorists" learn what substances are likely to produce certain flavors and aromas and how mixtures of substances can duplicate natural flavors. If you are wondering why it is necessary to add flavor to foods, think of all of the precooked, prepared foods you see in a supermarket—frozen foods you can heat and eat, food that has been processed to remove fat, instant coffee and tea, and foods such as soft drinks that don't even exist in a natural form. People want these sorts of products but expect them to taste good too. When foods are processed, many of their flavor and aroma components are lost or changed. So, flavor chemists analyze the components of the natural food and attempt to duplicate those flavors in the processed product. Sometimes chemists add flavors that don't even exist in the food's natural form.

Name _____ Class _____ Date _____

An Introduction to Food Chemistry *continued*

FIGURE 2: A COMPARISON OF FOUR VANILLOID COMPOUNDS

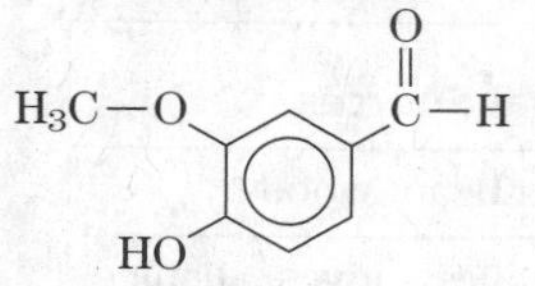

Vanillin

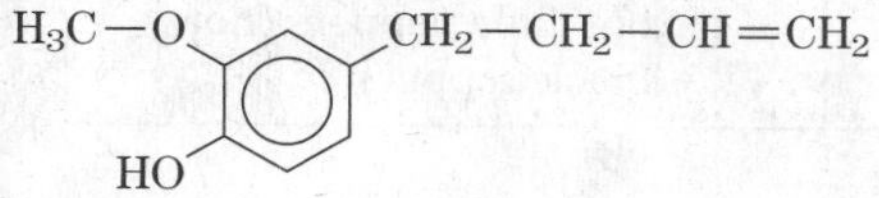

Eugenol

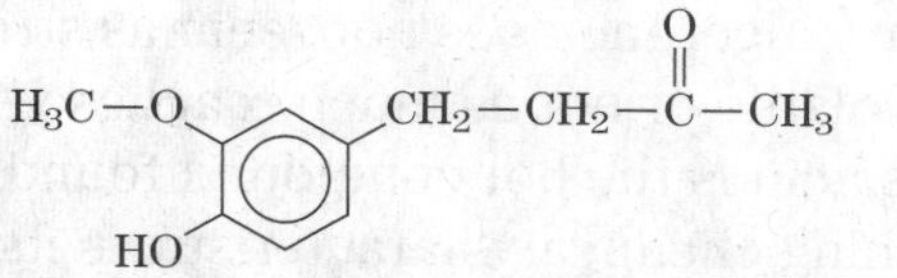

Zingerone

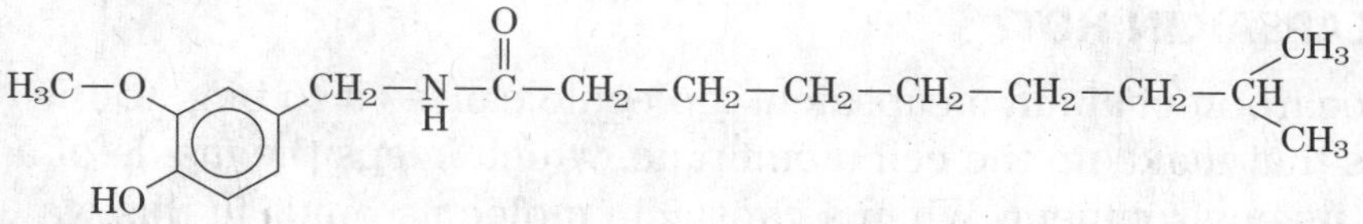

Capsaicin

FLAVOR AND MOLECULAR STRUCTURE

Scientists think that the flavor and aroma produced by a given molecule relates to the way the molecule interacts with a protein in a receptor connected to a nerve. However, chemists now know that flavor and aroma depend a lot on the shapes and sizes of molecules. Scientists continue to do research to pinpoint all of the features of molecules that contribute to flavor and aroma. Sometimes, small differences in molecular structure make a great difference in flavor. Look at the structural formulas of the four compounds shown in **Figure 2** above. These molecules are classified as *vanilloid* compounds because they have the same parent structure as vanillin. They all have distinctive odors, but each one smells very different from the others because of the differences in the "side chains" attached to the vanilloid parent structure.

FLAVORS OF THE VANILLOID FAMILY

Characteristics and sources of the vanilloid compounds shown in **Figure 2** are summarized in **Table 1** on the next page. Vanillin itself has the pleasant, sweet smell of vanilla. Eugenol, on the other hand, has a much stronger, spicy scent. In high concentrations, eugenol can actually numb pain, which is why oil of clove (which contains eugenol) has sometimes been used on toothaches. The strength

An Introduction to Food Chemistry *continued*

TABLE 1: A COMPARISON OF FOUR VANILLOID COMPOUNDS

Name	Flavor	Natural Sources
Vanillin	vanilla	vanilla bean, wood
Eugenol	clove	bay leaves, cloves, allspice
Zingerone	hot ginger	ginger root, mustard
Capsaicin	hot, burning taste	hot chile peppers (not black pepper)

of flavor sensation overwhelms the perception of the toothache pain. Zingerone is the main flavor in ginger. If you've ever taken a bite of raw ginger root, you know that zingerone has a pungent, hot flavor. Zingerone does not seem as strong as eugenol, however, because it is not as volatile or not as much reaches your smell receptors in gaseous form. Finally, capsaicin is the hot compound found in hot chile or cayenne peppers. Its distinguishing chemical characteristic is its long hydrocarbon side chain.

WHY IS CAPSAICIN HOT?

The hydrocarbon chain in a capsaicin molecule causes it to bind to the oily lipid molecules that make up the cell membrane, which keeps the inside of a cell separate from its environment. When a capsaicin molecule binds in this way, calcium ions flood into the cell. This action is one of the things that happens when cells are exposed to high temperatures. The result is that you perceive a sensation that is just like the experience of heat, which is why we call spicy food "hot."

The effect of capsaicin on cells is the reason that capsaicin is one of the components of pepper spray. Capsaicin produces a hot taste on the tongue, but capsaicin in the eyes and nasal passages can be extremely painful. Capsaicin is also useful in *counterirritant* lotions and creams. Counterirritants can relieve the sensation of pain by causing pain themselves when they are applied to the skin. Typically, they are applied to areas of the body with strains, sprains, and arthritic pain. Scientists are still trying to determine the exact mechanism by which counterirritants work. One theory is that substances such as capsaicin irritate the skin, causing pain that masks the muscle or joint pain. In addition, the skin irritation causes increased blood flow to the whole area, and this can help heal muscle or joint injury.

Food scientists are using both genetic engineering and classical breeding techniques to create new types of chiles to suit a variety of preferences.

Careers in Food Chemistry

Food chemists have opportunities to work in a wide variety of interesting careers. A few of these careers are described below.

FLAVORIST/PERFUMER

As you read above, flavor and aroma are almost inseparable. The main difference between flavorists and perfumers is that flavorists work with food and perfumers work with all kinds of scents, not just the ones used in perfumes. Flavorists participate in the development of food products and study the molecular changes that occur when food is cooked. Some flavorists do research to determine what substances contribute to the natural flavors of food and try to find ways to re-create those flavors.

Many perfumers work to find scents that will make consumer products, even ones unrelated to food, more attractive to customers. Think of the many household cleaning products and their huge variety of smells. Perfumers contribute to the scent of everything from soap to new cars. Today, most professional flavorists and perfumers hold college degrees, often in chemistry.

DIETICIAN

Dieticians must understand the chemistry of food and the reactions used to digest nutrients. Becoming a professional dietician requires a college degree, internship, and state licensing in most cases. Dieticians do more than just plan meals for schools and hospitals. Some dieticians work with professional, college, or Olympic sports teams as well as fitness and health clubs. Some dieticians are in private practice, where they prescribe diets for patients to ensure good nutrition while meeting some special need, such as gaining or losing weight, managing an allergy, or alleviating some diseases involving the digestive system.

FOOD TECHNOLOGIST

Food technologists deal with the composition and changes in food from agricultural production to processing, storage, and distribution. Students who want to be food technologists study chemistry and biology in college and also take specific courses in food technology. Some food technologists work in research to find ways to maintain acceptable tastes and textures in food as it is processed. Other food technologists work with engineers to develop processing, packaging, and shipping techniques that will ensure the quality of food products while meeting government regulations.

Name _____ Class _____ Date _____

An Introduction to Food Chemistry *continued*

Topic Questions

1. Why are flavoring substances added to processed foods?

Natural flavors and aromas are often changed or destroyed when food is

processed commercially. These flavors are restored by adding flavoring

substances.

2. Explain the difference between "taste" and "flavor."

The sense of taste recognizes only four or five fundamental tastes—sweet,

sour, bitter, salty, and perhaps a fifth taste called umami. Flavor includes not

only these fundamental tastes but also the aromas and chemical irritation

properties of food chemicals.

3. What structural feature of the vanilloid compounds is responsible for their very different flavors?

The composition and length of the side chain is responsible for the changes

in flavor.

4. Explain why chemical compounds such as eugenol and capsaicin can be used as anesthetics.

In the case of eugenol, its strong flavor gives a sensation that overrides pain

sensations. Capsaicin can be used externally on aching muscles as a coun-

terirritant: similarly to eugenol, the sensation of irritation it causes over-

whelms the sensation of muscle ache.

5. Draw the part of the chemical structure that is common to all four vanilloid compounds discussed.

Energy Content of Foods

Teacher Notes

TIME REQUIRED One 50-minute lab period (for a two-period lab, have each lab group test all four food samples)

LAB RATINGS

Easy ← 1 2 3 4 → Hard

Teacher Preparation–3
Student Setup–3
Concept Level–2
Cleanup–1

SKILLS ACQUIRED

Collecting data
Experimenting
Organizing and analyzing data
Interpreting
Drawing real-world conclusions

SCIENTIFIC METHODS

Analyze the Results In Analysis questions 1–5, students will compile the data from their experiments and make calculations to determine the caloric content of the foods tested.

Draw Conclusions In Conclusions questions 1–4, students will interpret the data and apply it to the objectives of the experiment.

MATERIALS (PER LAB GROUP)

- can, small
- food sample (2)
- food holder (see **Figure 1** on p. 155)
- graduated cylinder, 100 mL
- LabPro or CBL2 interface
- matches
- ring stand with a 4-in. ring

- slit stopper
- stirring rod (2)
- TI graphing calculator
- utility clamp
- Vernier temperature probe
- water, cold
- wooden splint

The food stand can be made using an extra-large paper clip and a small jar lid, such as a baby-food jar lid. Partially straighten the paper clip, then bend a small loop at one end. This loop will cradle food samples. Bend the other end to a V shape—this will be the base. Glue the paper clip into the lid. An advantage of such a stand is its ability to catch pieces of burned food that fall.

Small soup cans work well. Remove the paper and label the top. Place two holes, large enough to accommodate a stirring rod, near the top. Some teachers prefer to use aluminum beverage cans instead.

SAFETY CAUTIONS

Be sure to remove all sharp edges from cans.

Because peanuts and cashew nuts release very large amounts of energy as heat as they burn, you may want to have your students use 100 mL portions of cold water when testing these foods.

Some students may be allergic to peanuts. Before proceeding with this activity, poll your students to determine if anyone in the class is allergic to peanuts. If any are, do not allow any students to perform the part involving peanuts. Have students answer Conclusions question 2 for cashews instead of peanuts.

Graphing Calculator and Sensors
TIPS AND TRICKS

Students should have the DataMate program loaded on their graphing calculators. Refer to Appendix B of Vernier's *Chemistry with Calculators* for instructions.

Not all models of TI graphing calculators have the same amount of memory. If possible, instruct students to clear all calculator memory before loading the DataMate program.

The temperature calibrations that are stored in the DataMate data-collection program will work fine for this experiment. No calibration is necessary for the temperature probes.

The Vernier stainless steel temperature probe and CBL temperature probe will plug directly into CH1 on the Vernier LabPro or CBL2 interface. If you are using the Vernier direct-connect temperature probe, you will need a DIN-BTA (formerly CBL-DIN) adapter to convert from the 5-pin Din connector to the BTA connector.

NOTES ON TECHNIQUE

When viewing graphs on the calculator, students should use the arrow keys to trace the data points on the graph.

If students wish to see the data for both food samples on the same graph, instruct them to store the first data set before beginning the second food sample. From the Main Screen of DataMate, the Store Latest Run feature can be found under the Tools menu. The program will permit storing only up to two runs. If more than one sensor is used at a time, the Store Latest Run feature will not work.

Experimental Setup
TIPS AND TRICKS

Supply students with water that is 15°C to 18°C to achieve best results. Perform this experiment in a fume hood or in a well-ventilated classroom.

The following are corrected versions of the nutrition labels found on page 150.

MARSHMALLOWS (Corrected)

Nutrition Facts	
Serving Size 1 ounce	
Servings Per Container 6	
Amount per serving	
Calories 90	Calories from Fat 0
	% Daily Value
Total Fat 0g	*%
Saturated Fat 0g	*%
Cholesterol 0mg	*%
Sodium 13mg	*%
Total Carbohydrate 23g	8%
Dietary Fiber 0g	*%
Sugars 5.9g	
Protein 0.1g	
*Less than 1% of US RDA	

PEANUTS (oil roasted w/salt)

Nutrition Facts	
Serving Size 1 ounce	
Servings Per Container 16	
Amount per serving	
Calories 165	Calories from Fat 125
	% Daily Value
Total Fat 14g	70%
Saturated Fat 1.9g	35%
Cholesterol 0mg	0%
Sodium 122mg	18%
Total Carbohydrate 6g	6%
Dietary Fiber 3g	4%
Sugars 3g	
Protein 8g	

CASHEWS (oil roasted w/salt) (Corrected)

Nutrition Facts	
Serving Size 1 ounce	
Servings Per Container 16	
Amount per serving	
Calories 163	Calories from Fat 26
	% Daily Value
Total Fat 13.7g	69%
Saturated Fat 2.7g	48%
Cholesterol 0mg	0%
Sodium 177mg	26%
Total Carbohydrate 8g	10%
Dietary Fiber 1g	1%
Sugars 7g	
Protein 5g	

POPCORN (air-popped, no salt)

Nutrition Facts	
Serving Size 1 cup	
Servings Per Container 8	
Amount per serving	
Calories 30	Calories from Fat 0
	% Daily Value
Total Fat 0.3g	*%
Saturated Fat 0g	*%
Cholesterol 0mg	0%
Sodium 0mg	*%
Total Carbohydrate 6g	2%
Dietary Fiber 1g	4%
Sugars 2g	
Protein 0g	
*Less than 1% of US RDA	

DATA TABLES WITH SAMPLE DATA
DATA TABLE 1

Food sample 1:			
initial mass of food sample and holder:	14.04 g		
mass of empty can:	41.31 g	mass of can and water:	90.69 g

DATA TABLE 2

Food sample 1:			
T_1: 15.4°C	T_2: 52.9°C	final mass of sample and holder:	13.36

DATA TABLE 3

Food sample 1:			
mass of water heated:	49.38 g	temperature change, ΔT:	37.5 °C
mass of food burned:	0.68 g	heat, q:	7.74 kJ
energy content of food sample:			11.4 kJ/g

CLASS AVERAGES

Marshmallows	Peanuts	Cashews	Popcorn
5.2 kJ/g	11.8 kJ/g	11.5 kJ/g	6.7 kJ/g

Food Chemistry Lab

PROBEWARE LAB

Energy Content of Foods

You are a lab technician working for NASA. Recently you were given the job of deciding what type of foods should be included in the next space mission. Four food types have been selected as possible snacks for the astronauts. You need to determine which of these four food choices has the highest *energy content*, while adding the least amount of mass to the mission.

Your team will test two of the food types using a method known as calorimetry. During this process, you will burn a food sample positioned below a can containing a given amount of cold water. The water temperature will be monitored during the experiment using a temperature probe. By calculating the temperature change of the water, you will determine how much energy was released when the food sample burned.

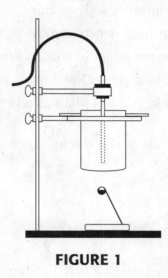

FIGURE 1

OBJECTIVES

Measure temperature changes.

Calculate energy changes using specific heat.

Infer the energy content of food.

Relate energy content to types of food.

Evaluate whether the nutrition labels are accurate.

MATERIALS

- can, small
- food sample (2)
- matches
- water, cold
- wooden splint

| **Energy Content of Foods** *continued*

EQUIPMENT

- food holder (see **Figure 1**)
- graduated cylinder, 100 mL
- LabPro or CBL2 interface
- stirring rod (2)
- ring stand with a 4-in. ring
- slit stopper
- TI graphing calculator
- utility clamp
- Vernier temperature probe

SAFETY

- Wear safety goggles when working around chemicals, acids, bases, flames, or heating devices. Contents under pressure may become projectiles and cause serious injury.

- Secure loose clothing, and remove dangling jewelry. Do not wear open-toed shoes or sandals in the lab.

- Wear an apron or lab coat to protect your clothing when working with chemicals.

- In order to avoid burns, wear heat-resistant gloves whenever instructed to do so.

- If you are unsure of whether an object is hot, do not touch it.

- Avoid wearing hair spray or hair gel on lab days.

- Whenever possible, use an electric hot plate as a heat source instead of an open flame.

- Never return unused chemicals to the original container; follow instructions for proper disposal.

Procedure

EQUIPMENT PREPARATION

1. Obtain and wear goggles.

2. Plug the temperature probe into Channel 1 of the LabPro or CBL2 interface. Use the link cable to connect the TI graphing calculator to the interface. Firmly press in the cable ends.

3. Turn on the calculator, and start the DATAMATE program. Press [CLEAR] to reset the program.

4. Set up the calculator and interface for the temperature probe.

   a. Select SETUP from the main screen.

   b. If the calculator displays a temperature probe in CH 1, proceed directly to Step 5. If it does not, continue with this step to set up your sensor manually.

   c. Press [ENTER] to select CH 1.

   d. Select TEMPERATURE from the SELECT SENSOR menu.

   e. Select the temperature probe you are using (in degrees Celsius) from the TEMPERATURE menu.

Energy Content of Foods *continued*

5. Set up the data-collection mode.

   a. To select MODE, press ▲ once and press ENTER.

   b. Select TIME GRAPH from the SELECT MODE menu.

   c. Select CHANGE TIME SETTINGS from the TIME GRAPH SETTINGS menu.

   d. Enter "6" as the time between samples in seconds.

   e. Enter "100" as the number of samples. The length of the data collection will be 10 minutes.

   f. Select OK to return to the setup screen.

   g. Select OK again to return to the main screen.

6. Obtain a piece of one of the two foods assigned to you and a food holder like the one shown in **Figure 1.** Find and record the initial mass of the food sample and food holder. **CAUTION:** *Do not eat or drink in the laboratory.*

7. Determine and record the mass of an empty can. Obtain cold water from your teacher, and add 50 mL of it to the can. Determine and record the mass of the can and water.

8. Set up the apparatus as shown in **Figure 1.** Use a ring and stirring rod to suspend the can about 2.5 cm (1 in.) above the food sample. Use a utility clamp to suspend the temperature probe in the water. The probe should not touch the bottom of the can. Remember that the temperature probe must be in the water for at least 30 seconds before you complete Step 9.

DATA TABLE 1

Food sample 1:			
initial mass of food sample and holder:			
mass of empty can:		mass of can and water:	
Food sample 2:			
initial mass of food sample and holder:			
mass of empty can:		mass of can and water:	

DATA COLLECTION

9. Select START to begin collecting data. Record the initial temperature of the water, T_1, in Data Table 2 (round to the nearest 0.1°C). **Note:** You can monitor temperature in the upper-right corner of the real-time graph displayed on the calculator screen. Remove the food sample from under the can, and use a wooden splint to light it. Quickly place the burning food sample directly under the center of the can. Allow the water to be heated until the food sample stops burning.

10. Continue stirring the water until the temperature stops rising. Record this maximum temperature, T_2. Data collection will stop after 10 minutes (or press the STO▸ key to stop *before* 10 minutes have elapsed).

11. Determine and record the final mass of the food sample and food holder.

12. To confirm the initial (T_1) and final (T_2) values you recorded earlier, examine the data points along the curve on the displayed graph. As you move the cursor right or left, the time (X) and temperature (Y) values of each data point are displayed below the graph.

13. Press [ENTER] to return to the main screen. Select START to repeat the data collection for the second food sample. Use a new 50 mL portion of cold water. Repeat Steps 6–12.

14. When you are done, place burned food, used matches, and partially burned wooden splints in the container provided by the teacher.

DATA TABLE 2

Food sample 1:					
T_1:		T_2:		final mass of sample and holder:	
Food sample 2:					
T_1:		T_2:		final mass of sample and holder:	

Analysis

1. **Organizing Data** Find the mass of water heated for each sample. _____

   Answers should be equal to mass of can and water – mass of empty can.

2. **Organizing Data** Find the change in temperature of the water, ΔT, for each sample. **Answers should be equal to $T_2 - T_1$.**

   _____

3. **Organizing Data** Find the mass (in grams) of each food sample burned. _____

   Answers should be equal to initial mass of food sample and holder – final mass.

4. **Analyzing Results** Calculate the heat absorbed by the water, q, using the equation

   $$q = C_p m \Delta T$$

   where q is heat, C_p is the specific heat, m is the mass of water, and ΔT is the change in temperature. For water, C_p is 4.18 J/g°C. Convert your final answer to units of kJ. **Answers may vary.**

   _____

   _____

Energy Content of Foods *continued*

5. Analyzing Results Use the results of the previous two steps to calculate the energy content (in kJ/g) of each food sample. _____

Students should divide their answers to #4 by their answer to #3.

_____

DATA TABLE 3

Food sample 1:					
mass of water heated:		g	temperature change, ΔT:		°C
mass of food burned:		g	heat, q:		kJ
energy content of food sample:					kJ/g
Food sample 2:					
mass of water heated:		g	temperature change, ΔT:		°C
mass of food burned:		g	heat, q:		kJ
energy content of food sample:					kJ/g

Conclusions

1. Evaluating Results Record your results and the results of other groups in the Class Results Table below. Which food had the highest energy content? Which had the lowest energy content? **Cashews and peanuts had the highest energy content. Marshmallows and popcorn had the lowest.**

CLASS RESULTS TABLE

Marshmallows	Peanuts	Cashews	Popcorn
kJ/g	kJ/g	kJ/g	kJ/g
kJ/g	kJ/g	kJ/g	kJ/g
kJ/g	kJ/g	kJ/g	kJ/g
kJ/g	kJ/g	kJ/g	kJ/g
kJ/g	kJ/g	kJ/g	kJ/g

Average for each food type:			
kJ/g	kJ/g	kJ/g	kJ/g

2. Evaluating Results Food energy is often expressed in a unit called a Calorie (or dietary calorie). There are 4.18 kJ in one Calorie. Based on the class average for peanuts, calculate the number of Calories in a 50.0 g package of peanuts. **Calories in a 50.0 g package of peanuts:**

(12.0 kJ/g)(50.0 g)(1 Cal/4.18 kJ) = 155 Cal

3. **Evaluating Results** Two of the foods in the experiment have a high fat content (peanuts and cashews), and two have a high carbohydrate content (marshmallows and popcorn). From your results, what generalization can you make about the relative energy content of fats and carbohydrates? _____

   The two foods with a high fat content, cashews and peanuts, have a much

   higher energy content than those with a high carbohydrate content.

4. **Evaluating Results** Based on the data you and your classmates collected, which of the four foods tested would you suggest to send on the NASA space mission?

   On average, peanuts have the highest energy content per gram, followed by

   cashews.

Extensions

1. **Applying Results** If you were packing for a mountain hike, what kind of snacks would you bring along? Why? _____

   Nuts of any kind would be a good energy source for the physical demands

   involved.

2. **Critiquing Methods** Was all of the energy as heat that was given off by the burning food sample transferred to the water in the can? How could this experiment be improved to account for all of the energy given off when the food sample was burned? _____

   Answers should discuss the possible loss of energy as heat between the

   burner and the thermometer and possible improvements to keep that loss

   near zero—for instance, insulating the space between the burner and the

   water to prevent heat loss.

3. **Applying Results** Listed on the following page are possible nutrition labels for each of the food samples that you tested. Based on the data you and your classmates obtained in this lab, determine which of these labels are accurate and which are not. If you find a label to be incorrect, explain your reasoning.

   The nutrition labels for peanuts and popcorn are accurate. The marshmallow

   label, however, indicates a higher Calorie and fat content than is likely, and

   the label for cashews indicates a fat and Calorie content that is too low.

   _____

Name _____ Class _____ Date _____

Energy Content of Foods continued

MARSHMALLOWS

Nutrition Facts

Serving Size 1 ounce	
Servings Per Container 6	
Amount per serving	
Calories 260	Calories from Fat 160
	% Daily Value
Total Fat 18g	**13%**
Saturated Fat 5g	**27%**
Cholesterol 0mg	**0%**
Sodium 260mg	**11%**
Total Carbohydrate 23g	**8%**
Dietary Fiber 1g	**11%**
Sugars 18g	
Protein 1g	

PEANUTS

Nutrition Facts

Serving Size 1 ounce	
Servings Per Container 16	
Amount per serving	
Calories 165	Calories from Fat 125
	% Daily Value
Total Fat 14g	**20%**
Saturated Fat 1.9g	**10%**
Cholesterol 0mg	**0%**
Sodium 122mg	**5%**
Total Carbohydrate 5g	**2%**
Dietary Fiber 1g	**4%**
Sugars 2g	
Protein 8g	

CASHEWS

Nutrition Facts

Serving Size 1 ounce	
Servings Per Container 16	
Amount per serving	
Calories 80	Calories from Fat 26
	% Daily Value
Total Fat 3g	**4%**
Saturated Fat 0.5g	**3%**
Cholesterol 0mg	**0%**
Sodium 177mg	**7%**
Total Carbohydrate 8g	**3%**
Dietary Fiber 2g	**8%**
Sugars 2g	
Protein 5g	

POPCORN

Nutrition Facts

Serving Size 1 cup	
Servings Per Container 8	
Amount per serving	
Calories 30	Calories from Fat 0
	% Daily Value
Total Fat 0.3g	*%
Saturated Fat 0g	*%
Cholesterol 0mg	**0%**
Sodium 0mg	*%
Total Carbohydrate 6g	**2%**
Dietary Fiber 1g	**4%**
Sugars 2g	
Protein 0g	

*Less than 1% of US RDA

PROBEWARE LAB

Buffer Capacity in Commercial Beverages

Teacher Notes

TIME REQUIRED One 50-minute lab period (extension may require additional time)

LAB RATINGS

Easy ←—— 1 　2 　3 　4 ——→ Hard

Teacher Preparation–4
Student Setup–3
Concept Level–4
Cleanup–3

SKILLS ACQUIRED

Collecting data
Experimenting
Organizing and analyzing data
Interpreting
Drawing real-world conclusions

SCIENTIFIC METHODS

Make Observations Students will collect pH and volume data using a calculator-interfaced pH sensor.

Analyze the Results In Analysis questions 1 and 2, students will analyze the data from their experiments and identify patterns.

Draw Conclusions In Conclusions questions 1–3, students will interpret the data and relate their results to the real-world example given in the introduction.

MATERIALS (PER LAB GROUP)

- beaker, 250 mL
- citric acid solution, 0.01 M, 40 mL
- graduated cylinder, 50 mL or 100 mL
- LabPro or CBL2 interface
- lemonade drink, 40 mL
- NaOH solution, 0.01 M, 100 mL
- magnetic stirrer with stirring bar
- ring stand
- TI graphing calculator
- utility clamp (2)
- Vernier pH sensor
- rinse bottle with distilled water
- water, distilled, about 120 mL

Prepare 2 qt of the lemonade drink using tap water. There are several brands of drink mix that use a citric acid–sodium citrate buffer. The sample data is based on Country Time® Lemonade Mix, which works quite well.

Prepare the 0.010 M citric acid solution by dissolving 1.92 g of $H_3C_6H_5O_7$ (or 2.10 g of $H_3C_6H_5O_7 \cdot H_2O$) in 1.00 L of solution.

The 0.1 M NaOH solution can be prepared by adding 4.0 g of NaOH to make 1 L of solution.

SAFETY CAUTIONS

Sodium hydroxide is a corrosive solid and is known to cause skin burns upon contact. Wear gloves when working with this substance to prepare solutions. When this solution is added to water, much energy as heat is evolved. This is very dangerous to eyes, so wear face and eye protection when using this substance. When using chemicals, students should wear aprons, gloves, and goggles.

Graphing Calculator and Sensors

TIPS AND TRICKS

Students should have the DataMate program loaded on their graphing calculators. Refer to Appendix B of Vernier's *Chemistry with Calculators* for instructions. The pH calibrations that are stored in the DataMate data-collection program will work fine for this experiment. For more accurate pH readings, you (or your students) can do a 2-point calibration for each pH system using pH 4 and pH 7 buffers.

An alternate way of determining the precise equivalence point of the titration is to take the second derivative of the pH-volume data. When DataMate is transferred to the calculator, a small program called PHDERIVS will be copied into the calculator as well (PHDERIVS will not be loaded onto TI-73 and TI-83 calculators because of their memory capabilities. If you would like a copy of PHDERIVS for these calculators, visit the Vernier Software & Technology Web site at www.vernier.com/cbl/progs.html). PHDERIVS allows you to view first and second derivative plots of pH-volume data. PHDERIVS is set up to analyze volume data in L1 and pH data in L2. To run the program, follow this procedure:

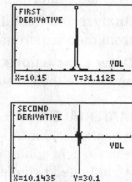

a. After you have collected pH-volume data, leave DataMate by selecting QUIT from the main screen.

b. Start the PHDERIVS program. (Note: On a TI-83 Plus, PHDERIVS will be loaded as a program [press PRGM, not APPS]. On a TI-86, 89, 92, or 92 Plus, the PHDERIVS program will be listed alphabetically below the DATAMATE programs.)

c. Proceed to the GRAPHS menu. Select SECOND DERIV, a plot of $\Delta^2 pH / \Delta vol^2$.

d. Using the arrow keys, move the cursor to the equivalence point. This will be the point where the curve crosses the zero line. The x-value shown at the bottom of the screen is the volume of acid at the equivalence point.

NOTES ON TECHNIQUE

Before students view graphs on the calculator, remind them how to use the arrow keys to trace the data points on the graph.

Experimental Setup
TIPS AND TRICKS

Consider dispensing the 40 mL of lemonade drink mix and citric acid solution required for each lab team from a buret instead of a pipet. Consider having your students add two or three drops of phenolphthalein indicator at the beginning of each titration. They can then observe the phenolphthalein equivalence point and compare it with the pH equivalence point for the titration.

Lab teams should consist of two or three students. The titration will proceed faster if one student operates the buret while another student manipulates the calculator.

Answers
DATA TABLES WITH SAMPLE DATA

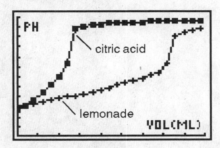

Titration of buffered drink mix and 0.01 M citric acid using 0.10 M NaOH

DATA TABLE

	Lemonade	Citric acid solution
Concentration of NaOH	0.10 M	0.10 M
Volume of NaOH added *before* largest pH change	36.0 mL	12.0 mL
Volume of NaOH added *after* largest pH change	38.0 mL	14.0 mL
Volume of NaOH added at equivalence point	37.0 mL	13.0 mL
Mole NaOH	0.00370 mol	0.00130 mol

Buffer Capacity in Commercial Beverages

Quickslurp Beverages, Inc., has just produced a new lemonade drink mix. As a food scientist working in the quality control department, your job is to ensure that the product tastes the same no matter what type of water is used to prepare it. The difference in pH of varying water samples can affect the overall taste of the drink. To prevent any change in taste, a citric acid–sodium citrate buffer has been added to the mix. You are responsible for ensuring that this buffer system properly resists small changes in pH that may result when water is added.

Household tap water can vary in pH from 6.5 to 8.5. That means that some water sources are slightly acidic, whereas others are slightly basic. The buffer added to the drink mix can resist changes in pH upon the addition of small amounts of H^+ or OH^- ions, preventing any change in the drink's taste. Combining a weak acid with a salt of the weak acid can form one type of buffer. Citric acid and sodium citrate have been added to the drink mix and are an example of this kind of buffer pair.

You will make a direct comparison of buffer capacity between the lemonade drink mix and an unbuffered 0.010 M citric acid solution. The pH of the drink mix will be monitored with a pH sensor as a solution of 0.10 M NaOH is titrated into the solution. A second titration will be performed using the citric acid solution in place of the drink mix. A comparison between the two titrations will reveal the buffering capacity of the lemonade drink mix.

OBJECTIVES

Measure pH changes.

Graph pH-volume data pairs.

Compare pH change in buffered and unbuffered solutions.

Calculate buffer capacity of a lemonade drink.

MATERIALS

- citric acid solution, 0.01 M
- lemonade drink
- NaOH solution, 0.10 M
- rinse bottle with distilled water
- water, distilled

Name _____ Class _____ Date _____

Buffer Capacity in Commercial Beverages *continued*

EQUIPMENT

- beaker, 250 mL
- graduated cylinder, 50 mL or 100 mL
- LabPro or CBL2 interface
- magnetic stirrer with stirring bar (if available)

- ring stand
- TI graphing calculator
- utility clamp (2)
- Vernier pH sensor

SAFETY

- Wear safety goggles when working around chemicals, acids, bases, flames, or heating devices. Contents under pressure may become projectiles and cause serious injury.

- If any substance gets in your eyes, notify your instructor immediately and flush your eyes with running water for at least 15 minutes.

- If a chemical is spilled on the floor or lab bench, alert your instructor, but do not clean it up yourself unless your teacher says it is OK to do so.

- Secure loose clothing and remove dangling jewelry. Do not wear open-toed shoes or sandals in the lab.

- Wear an apron or lab coat to protect your clothing when working with chemicals.

- Never return unused chemicals to the original container; follow instructions for proper disposal.

- Always use caution when working with chemicals.

- Never mix chemicals unless specifically directed to do so.

- Never taste, touch, or smell chemicals unless specifically directed to do so.

Procedure

EQUIPMENT PREPARATION

1. Obtain and wear goggles.

2. Use a graduated cylinder to measure out 40 mL of the lemonade drink and 60 mL of distilled water into a 250 mL beaker. **CAUTION:** *Do not eat or drink in the laboratory.*

3. Place the beaker on a magnetic stirrer and add a stirring bar. If no magnetic stirrer is available, you will need to stir the beaker with a stirring rod during the titration.

4. Plug the pH sensor into Channel 1 of the LabPro or CBL 2 interface. Use the link cable to connect the TI graphing calculator to the interface. Firmly press in the cable ends.

Buffer Capacity in Commercial Beverages *continued*

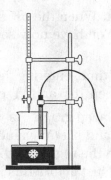

FIGURE 1

5. Use a utility clamp to suspend a pH sensor on a ring stand as shown in **Figure 1**. Position the pH sensor in the lemonade mixture, and adjust its position so that it is not struck by the stirring bar.

6. Obtain a 50 mL buret, and rinse it with a few mL of the 0.10 M NaOH solution. **CAUTION:** *Sodium hydroxide solution is caustic. Avoid spilling it on your skin or clothing.* Dispose of the rinse solution as directed by your teacher. Use a utility clamp to attach the buret to the ring stand as shown in **Figure 1**. Fill the buret a little above the 0.00 mL level with 0.10 M NaOH solution. Drain a small amount of NaOH solution so that it fills the buret tip *and* leaves the NaOH at the 0.00 mL level of the buret. Record the precise concentration of the NaOH solution in your data table.

DATA COLLECTION

7. Turn on the calculator, and start the DATAMATE program. Press [CLEAR] to reset the program.

8. Set up the calculator and interface for the pH sensor.

   a. Select SETUP from the main screen.

   b. If CH 1 displays PH, proceed directly to Step 9. If it does not, continue with this step to set up your sensor manually.

   c. Press [ENTER] to select CH 1.

   d. Select PH from the SELECT SENSOR menu.

9. Set up the data-collection mode.

   a. To select MODE, press [▲] once and press [ENTER].

   b. Select EVENTS WITH ENTRY from the SELECT MODE menu.

   c. Select OK to return to the main screen.

10. You are now ready to perform the titration. This process goes faster if one person adjusts and reads the buret while another person operates the calculator and enters volume data.

   a. Select START to begin data collection.

   b. Before you have added any NaOH solution, press [ENTER] and type in "0" as the buret volume in mL. Press [ENTER] to save the first data pair for this experiment.

Buffer Capacity in Commercial Beverages *continued*

c. Add 2.0 mL of NaOH titrant. When the pH stabilizes, press [ENTER] and enter the current buret reading. You have now saved the second data pair for the experiment.

d. Continue to add 2.0 mL at a time, entering the buret level after each one. When the pH has leveled off between 10.5 and 11, press [ST●►] to end data collection.

11. Examine the data on the displayed graph. As you move the cursor right or left on the graph, the volume (X) and pH (Y) values of each data point are displayed below the graph. Go to the region of the graph with the largest increase in pH. Find the NaOH volume just *before* this jump. Record this value in the data table. Then record the NaOH volume *after* the 2 mL addition producing the largest pH increase.

12. Store the data from the first run so that it can be used later:

a. Press [ENTER] to return to the main screen, and then select TOOLS.

b. Select STORE LATEST RUN from the TOOLS menu.

13. Use a graduated cylinder to measure out 40 mL of 0.010 M citric acid solution and 60 mL of distilled water into a 250 mL beaker. Position the pH sensor, beaker, and stirring bar as you did in Step 5. Refill the buret to the 0.00 mL level of the buret with 0.10 M NaOH solution. **CAUTION:** *Sodium hydroxide solution is caustic. Avoid spilling it on your skin or clothing.*

14. Repeat Steps 6 and 10–11 of the procedure. **Important:** Add the same total volume of NaOH that you did in the first trial using the same number of 2 mL additions. Both lists of data must have the same number of points and the same volumes to be compatible for graphing in Step 16.

15. When you are finished, dispose of the beaker contents as directed by your teacher. Rinse the pH sensor, and return it to the pH storage solution.

16. A good way to compare the two curves is to view both sets of data on one graph:

a. Press [ENTER] to return to the main screen.

b. Select GRAPH from the main screen, then press [ENTER].

c. Select MORE, then select L2 AND L3 VS L1 from the MORE OPTIONS menu.

d. Both pH runs should now be displayed on the same graph. Each point of the first run (lemonade drink) is plotted with a box, and each point of the second run (unbuffered citric acid solution) is plotted with a dot.

17. Print a graph of pH versus volume (with two curves displayed). Label each curve as "buffered lemonade" or "unbuffered citric acid."

Buffer Capacity in Commercial Beverages *continued*

DATA TABLE

	Lemonade	Citric acid solution
Concentration of NaOH	M	M
Volume of NaOH added *before* largest pH change	mL	mL
Volume of NaOH added *after* largest pH change	mL	mL
Volume of NaOH added at equivalence point	mL	mL
Mole NaOH	mol	mol

Analysis

1. **Examining Data** Use your graph and data table to locate the equivalence point. The equivalence point is characterized as the region of greatest pH change. Add the two NaOH values determined in Step 11, and divide by two. Record the result in your data table as the volume of NaOH added at the equivalence point.

2. **Organizing Data** Using the volume of NaOH titrant added, calculate the number of moles of NaOH used in each titration. Record the results in your data table.

Conclusions

1. **Interpreting Information** Which of the two solutions required more NaOH to reach the equivalence point? <u>The lemonade drink mix required significantly</u> <u>more NaOH to reach the equivalence point.</u>

2. **Analyzing Graphs** Compare the graph of each titration. How does the titration curve of the buffered lemonade drink mix compare with that of the unbuffered citric acid solution? <u>The graph of the citric acid titration shows a</u> <u>steep rise in pH from the first moment the base is added. The titration curve</u> <u>begins to level out at a pH above 10 after the addition of 14 mL of NaOH.</u> <u>The graph of the lemonade drink mix titration shows a slight and steady rise</u> <u>in pH as the base is added. The titration curve begins to level out at a pH</u> <u>above 10 after the addition of 38 mL of NaOH.</u>

3. **Applying Conclusions** Based on the results of this experiment, do you believe the buffer system of the prepared lemonade drink is sufficient to handle more extreme pH changes? **The buffering capacity of the drink mix appears to work well enough to sufficiently handle small changes in pH, but not large ones.**

Extensions

1. **Designing Experiments** Prepare three equal volume samples of lemonade drink using water samples with a pH of 6, 7, and 8. Test the resulting pH of the lemonade drink by using the pH sensor.

Name _____ Class _____ Date _____

Topic Introduction

An Introduction to Chemical Engineering

An engineer is a person who finds ways to put scientific knowledge to practical use. **Chemical engineers** *are people who study processes that involve matter and find ways to monitor and control those processes, often on an industrial scale.* These processes can be anything from the transport of solids, liquids, or gases, to carrying out series of chemical reactions on an industrial scale. For example, a chemical engineer may use a knowledge of chemistry to design ways of producing chemical products including pharmaceuticals, paint, petroleum products, plastics and other polymers, paper, fertilizers, and even the materials used to make microchips.

Chemical engineers work in many other areas. Some chemical engineers work in environmental fields to find ways to control and eliminate pollution. Other chemical engineers work in medicine and biotechnology, where they may design devices that control the flow of gases and liquids, improve kidney dialysis machines, or find new materials that can be used for medical devices within the body. Think of all the complex devices you might find in the operating rooms and patient care facilities of a large hospital. Someone had to design all of these items, and a chemical engineer was probably involved.

A Current Hot Topic: Biomedical Engineering

Biomedical engineering *is one of the newest branches of engineering and combines chemical engineering, biology, medicine, and computer science to devise new ways to improve health care.* It is such a growing field in today's society that chemical engineering departments at many universities have changed their names to include biomedical engineering.

Biomedical engineers work with physicians, surgeons, dentists, nurses, technicians, and researchers in a wide range of tasks. For example, biomedical engineers may develop new processes for diagnosis and treatment, research new types of materials that can take the place of human tissues or speed up healing, or design improved prosthetic devices that can move in response to nerve impulses.

BIOINSTRUMENTATION

In the area of **bioinstrumentation,** *a biomedical engineer works to find ways to control electronic devices used in diagnosis, treatment, and patient monitoring with computers or to record and analyze the data that the devices gather.* A bioinstrumentation engineer may devise or improve ways to have computers analyze the information from imaging techniques such as magnetic resonance imaging (MRI) scans. An MRI device is shown in **Figure 1** on the next page.

Name _____ Class _____ Date _____

An Introduction to Chemical Engineering *continued*

FIGURE 1: MAGNETIC RESONANCE IMAGING

Inside an MRI chamber, an extremely strong magnetic field is applied to the body. Hydrogen atoms in the body's molecules resonate in response and emit radio signals. These signals are analyzed by a computer program that constructs an image of a cross section of the body.

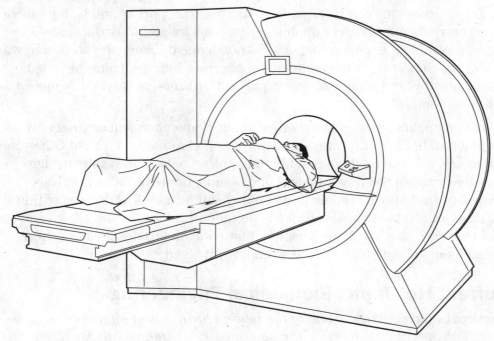

BIOMECHANICS

All engineers study mechanics, the science of forces and motion. Biomedical engineers use *biomechanics* to analyze motion within the body, including the interaction of muscles, tendons, ligaments, and bone. Also, biomechanical engineers study the motion of bodily fluids, such as blood in the circulatory and respiratory systems. Advances in biomechanics have increased understanding of how the heart, lungs, blood vessels, and capillaries work. Knowledge of biomechanics has led to the development and improvement of devices such as replacement joints, artificial hearts and heart valves, and artificial kidneys.

ORTHOPEDIC AND REHABILITATION BIOENGINEERING

Orthopedic and rehabilitation bioengineering is closely related to biomechanics. Engineers in this field may develop new materials to use in replacement joints so that the joints behave like natural joints. Research is underway to find materials that can take the place of tendons, ligaments, and cartilage. Some bioengineers work in the field of sports medicine, devising ways in which athletes can use their energy most efficiently and designing methods and devices to treat sports injuries.

Engineers are also working on devices to overcome sensory loss. An example is an artificial retina to enable people with failed retinas to see. In one approach, a tiny video camera is mounted on eyeglasses. The video signal is sent to a microchip on the retina where it stimulates some of the remaining undamaged retinal cells. The image produced is not very clear, but it enables the person to distinguish faces and many objects. Other engineers are developing an artificial retina that can send signals to the optic nerve or directly to the brain and can form more-detailed images.

BIOMATERIALS ENGINEERING

Biomaterials engineers develop materials that can safely be used in the human body. Materials to be used in the body must be unreactive, nontoxic, and sturdy, whether they are to be left in the body permanently or are in the body briefly, during surgery. One of these materials you may be familiar with is *cyanoacrylate* (super glue), which is often used to close wounds and surgical incisions. When glue is used, no stitches need to be removed later.

One focus of research in biomaterials is to develop "scaffold" materials that can act as support for the regrowth of tissue such as new skin for burn patients or new bone for patients with severe injuries or that have had bones removed because of cancer. These scaffolds can be made of ceramics, plastics, or other biological materials. Natural coral is sometimes used as a support for the regrowth of bone tissue.

A goal of engineers in biomaterials is to find a way to allow the body to regrow neurons. Such a discovery could help people who are paralyzed as a result of spinal cord injuries. Other research in biomaterials is directed at finding ways to regrow missing body parts. Before that can happen, scientists must learn more about the genetic factors that cause tissue to form certain structures such as bones, muscles, connective tissue, and skin.

Careers in Chemical Engineering
MEDICAL IMAGING TECHNOLOGIST

Medical imaging involves many techniques that have been developed to produce images of the internal structure of the body. You probably already know about X ray imaging, in which high-energy electromagnetic radiation passes through the body but is partially or completely blocked by internal structures. As the rays exit the body, they expose a piece of photographic film. When the film is developed, a "shadowgram" of the internal organs is revealed.

Ultrasound imaging is another technique in which high-frequency sound waves reflect from internal structures and the echoes are used to produce an image. Other techniques include positron emission tomography (PET), in which a small amount of a radioactive isotope is injected into the body. The radiation from the isotope is analyzed by a computer and an image is created. A medical imaging technician may prepare a patient, explain the imaging procedure, operate the imaging equipment, and make sure that the resulting images clearly show what the doctor wishes to see.

PILOT PLANT ENGINEER/TECHNICIAN

When a company decides to produce a new product, engineers often try out the manufacturing process on a smaller scale, using what is called a *pilot plant*. Chemical engineers are usually involved when the process involves combinations of matter or chemical reactions to produce new substances such as drugs. A pilot plant technician works with the engineers to find a way to obtain a good quality product as economically as possible. If the process involves chemical reactions, engineers use the pilot plant to determine the temperatures, pressures, flow rates, concentrations, and catalysts that give the best reaction rates and product yields. The technician may measure chemicals, record pilot plant data, control the process, use computers to analyze process data, and report results to the engineering staff.

ADVANCED MATERIALS ENGINEER

Materials engineers are always on the lookout for new materials with better properties. Advanced materials engineers develop new materials to meet specific needs. For example, engineers in the automobile and aerospace industries are always interested in making their vehicles lighter without sacrificing strength. Engineers may want a metal alloy (a mixture of two or more metals) that is light yet extremely strong. Some of the newest materials are those used in nanotechnology, which deals with the development of complex machines so small that they can be seen only with electron microscopes.

Name _____ Class _____ Date _____

An Introduction to Chemical Engineering *continued*

Topic Questions

1. What does a chemical engineer do?

A chemical engineer works with processes that involve matter and finds ways to monitor and control those processes.

2. Name at least three other kinds of professionals with which a biomedical engineer might work.

Biomedical engineers work with other medical personnel such as physicians, surgeons, dentists, nurses, and technicians as well as with research scientists and other engineers.

3. Describe a method that has been developed in biomedical engineering that enables visually impaired people to see.

Artificial retinas have been developed, as in the example of a tiny video camera mounted on eyeglasses, where the video signal is sent to a chip that stimulates the visual cells on the retina. In a variation on this technique, an artificial retina sends signals to the optic nerve, or directly to the brain, in order to form better images.

4. What is the function of scaffold materials in biomedical engineering?

Scaffold materials serve as support for the regrowth of natural tissues such as skin or bone.

5. How might a chemical engineer contribute to the development of an airplane that is lighter yet stronger than airplanes used today?

In advanced materials engineering, an engineer might develop new metal alloys that are lighter yet stronger than metals already in use in airplanes today.

Micro-Voltaic Cells

Teacher Notes

TIME REQUIRED One 50-minute lab period

LAB RATINGS Easy ←— 1   2   3   4 —→ Hard

    Teacher Preparation–4
    Student Setup–3
    Concept Level–3
    Cleanup–3

SKILLS ACQUIRED
    Collecting data
    Experimenting
    Organizing and analyzing data
    Interpreting
    Identifying/recognizing patterns

SCIENTIFIC METHODS

Make Observations Students will measure potential differences between half-cell pairs using a calculator-interfaced voltage probe.

Form a Hypothesis Students will calculate and predict reduction potentials for each half-cell pair.

Analyze the Results Students will record data and establish a table of reduction potentials.

Draw Conclusions Students will compare predicted reduction potentials with collected reduction potentials and calculate percentage error.

MATERIALS (PER LAB GROUP)

- forceps
- glass plate, 15 cm × 15 cm, or Petri dish, 11.5 cm diameter
- LabPro or CBL2 interface
- metals M_1, M_2, M_3, M_4, and M_5, 1 cm × 1 cm each (see next page)
- $NaNO_3$, 1 M
- paper, filter, 11.0 cm diameter
- sandpaper
- solutions of M_1^{2+}, M_2^{2+}, . . . , and M_5^{2+}, 1 M each (see below)
- TI graphing calculator
- voltage probe

Listed below are the quantities needed to prepare each of the solutions required for this experiment. Use distilled water for all solutions. The amounts shown are sufficient to fill three sets of 30 mL dropper bottles. This provides convenient dispensing for a class of 25 students.

1.0 M $CuSO_4$ (M_1^{2+}) (24.96 g solid $CuSO_4 \cdot 5H_2O$ per 100 mL)
1.0 M $ZnSO_4$ (M_2^{2+}) (26.95 g solid $ZnSO_4 \cdot 7H_2O$ per 100 mL)
1.0 M $Pb(NO_3)_2$ (M_3^{2+}) (33.10 g solid $Pb(NO_3)_2$ per 100 mL)
1.0 M $AgNO_3$ (M_4^+) (16.99 g solid $AgNO_3$ per 100 mL)
1.0 M $FeSO_4$ (M_5^{2+}) (27.80 g solid $FeSO_4 \cdot 7H_2O$ per 100 mL)
1.0 M $NaNO_3$ (8.50 g solid $NaNO_3$ per 100 mL)

The micro-scale dimensions of this lab provide tremendous savings in quantities of reagents used in preparing 1.0 M solutions. Each student uses approximately 0.15 mL of each solution in preparing the voltaic cells; thus, the total amount of solution used for a class is less than 5 mL. A 30 mL dropper bottle of solution can be expected to last for several years. The micro-scale dimensions also allow you to use silver metal and 1.0 M $AgNO_3$ solution. The cost of such a macro-scale cell would normally be prohibitive.

All of these solutions will store well except the 1.0 M $FeSO_4$. It is best to prepare this solution fresh each year. Store the 1.0 M $AgNO_3$ solution in an opaque or brown glass bottle. It should also be kept in a dark cabinet when not in use.

Each of the metal pieces should be cut in sizes approximately 1 cm $\times$ 1 cm. Cut each of the metals into distinct shapes (square, triangle, trapezoid) in order to aid students in returning the proper metal to the proper container. The metals can be coded as follows:

$$M_1 = Cu; \ M_2 = Zn; \ M_3 = Pb; \ M_4 = Ag; \ M_5 = Fe$$

Graphing Calculator and Sensors
TIPS AND TRICKS

Students should have the DataMate program loaded on their graphing calculators. Refer to Appendix B of Vernier's *Chemistry with Calculators* for instructions. The voltage calibration stored in the DataMate data-collection program will work fine for this experiment.

Not all models of TI graphing calculators have the same amount of memory. If possible, instruct students to clear all calculator memory before loading the DataMate program.

Answers
ANALYSIS

3. The identity of each of the metals is listed below along with the measured reduction potentials and the accepted reduction potentials from the Table of Standard Reduction Potentials. Most of the metals should easily be identified with the exception of iron, which may be mistaken for nickel.

Metal	Experiment reduction potential (E°)	Standard reduction potential (E°)
M_2 (Zn)	−0.74 V	−0.76 V
M_5 (Fe)	−0.30 V	−0.44 V
M_3 (Pb)	−0.13 V	−0.13 V
M_1 (Cu)	+0.34 V	+0.34 V
M_4 (Ag)	+0.79 V	+0.80 V

DATA TABLES WITH SAMPLE DATA
DATA TABLE 1

Voltaic cell (metals used)	Measured potential (V)	Metal number of (+) lead	Metal number of (−) lead
M_1/M_2	1.08 V	M_1	M_2
M_1/M_3	0.47 V	M_1	M_3
M_1/M_4	0.45 V	M_4	M_1
M_1/M_5	0.64 V	M_1	M_5

DATA TABLE 2

	Predicted potential (V)	Measured potential (V)	Percentage error (%)
M_2/M_3	−0.47 − (−1.08) = 0.61 V	0.60 V	1.6 %
M_2/M_4	0.45 − (−1.08) = 1.53 V	1.49 V	2.6 %
M_2/M_5	−0.64 − (−1.08) = 0.44 V	0.46 V	4.5 %
M_3/M_4	0.45 − (−0.47) = 0.92 V	0.90 V	2.2 %
M_3/M_5	−0.47 − (−0.64) = 0.17 V	0.16 V	5.9 %
M_4/M_5	0.45 − (−0.64) = 1.09 V	1.07 V	1.8 %

DATA TABLE 3

Metal (M_x)	Lowest (−) reduction potential, E° (V)
M_2 (Zn/Zn^{2+})	−1.08 V
M_5 (Fe/Fe^{2+})	−0.64 V
M_3 (Pb/Pb^{2+})	−0.47 V
M_1 (Cu/Cu^{2+})	0.00 V
M_4 (Ag/Ag$^+$)	+0.45 V
	Highest (+) reduction potential, E° (V)

Chemical Engineering Lab

Micro-Voltaic Cells

Your small plane has just crashed on a remote island in the South Pacific. After recovering from the crash, you inventory the plane's contents. The only useful item still working in the plane is the radio, but the plane's battery has been completely destroyed, and you have no other means of powering the radio. Wandering around the island, you discover a small building that appears to be an abandoned research facility. Inside you discover a working gasoline generator and a laboratory filled with chemicals and various items of lab equipment, but no batteries. Upon careful consideration, you decide that there is no way to get the generator to the radio or the radio to the lab. You decide that your best bet is to construct a new battery.

In one of the cabinets, you discover a collection of metal strips and solutions. The writing is in a foreign language with which you are unfamiliar. It is apparent from the writing that certain metals correspond to certain solutions. Remembering a little about electrochemistry from your high school chemistry class, you decide that you can use the metal strips and solutions to build an electrochemical cell. You realize that before you can begin, you need to establish a table of reduction potentials in order to choose the proper metals for the anode and cathode of the cell. After further searching, you discover a working voltmeter and all the materials necessary to create a series of micro-voltaic cells using the unknown metals and solutions.

In this experiment, you will be using a calculator-interfaced voltage probe in place of a voltmeter. The (+) lead makes contact with one metal and the (−) lead with another. If a positive voltage appears on the calculator screen, the cell has been connected correctly. If the voltage reading is negative, switch the positions of the leads. The metal attached to the (+) lead is the cathode (where reduction takes place) and thus has a higher, more positive reduction potential. The metal attached to the (−) lead is the anode (where oxidation takes place) and has the lower, more negative reduction potential.

OBJECTIVES

Measure potential differences between various pairs of half-cells.

Predict potentials of half-cell combinations.

Compare measured cell potentials with predicted cell potentials.

Calculate percentage error for measured potentials.

Establish the reduction potentials for five unknown metals.

Name _____ Class _____ Date _____

Micro-Voltaic Cells *continued*

MATERIALS

- filter paper, 11.0 cm diameter
- metals M_1, M_2, M_3, M_4, and M_5, 1 cm × 1 cm each
- $NaNO_3$, 1 M
- sandpaper
- solutions of M_1^{2+}, M_2^{2+}, . . . , and M_5^{2+}, 1 M each

EQUIPMENT

- forceps
- glass plate, 15 cm × 15 cm, or Petri dish, 11.5 cm diameter
- LabPro or CBL2 interface
- TI graphing calculator
- voltage probe

SAFETY

- Wear safety goggles when working around chemicals, acids, bases, flames, or heating devices.

- If any substance gets in your eyes, notify your instructor immediately and flush your eyes with running water for at least 15 minutes.

- If a chemical is spilled on the floor or lab bench, alert your instructor, but do not clean it up unless your instructor says it is OK to do so.

- Secure loose clothing, and remove dangling jewelry. Do not wear open-toed shoes or sandals in the lab.

- Wear an apron or lab coat to protect your clothing when working with chemicals.

- Never return unused chemicals to the original container; follow instructions for proper disposal.

- Always use caution when working with chemicals.

- Never mix chemicals unless specifically directed to do so.

- Never taste, touch, or smell chemicals unless specifically directed to do so.

Procedure

EQUIPMENT PREPARATION

1. Obtain and wear goggles.

2. Plug the voltage probe into Channel 1 of the LabPro or CBL2 interface. Use the link cable to connect the TI graphing calculator to the interface. Firmly press in the cable ends.

3. Turn on the calculator, and start the DATAMATE program. Press CLEAR to reset the program.

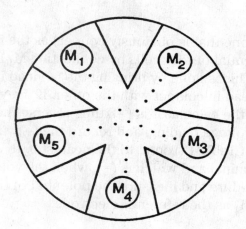

FIGURE 1

4. Set up the calculator and interface for the voltage probe.

  a. If the calculator displays VOLTAGE (V) in CH 1, proceed directly to Step 5. If it does not, continue with this step to set up your sensor manually.

  b. Select SETUP from the main screen.

  c. Press ⏍ENTER⏎ to select CH 1.

  d. Select the voltage probe you are using from the SELECT SENSOR menu.

  e. Select OK to return to the main screen.

5. Obtain a piece of filter paper, and draw five small circles with connecting lines, as shown in **Figure 1.** Using a pair of scissors, cut wedges between the circles as shown. Label the circles "M_1," "M_2," "M_3," "M_4," and "M_5." Place the filter paper on top of the glass plate.

6. Obtain five pieces of metal, M_1, M_2, M_3, M_4, and M_5. Sand each piece of metal on both sides. Place each metal near the circle with the same number.

7. Place three drops of each solution on its circle (M_1^{2+} on M_1, etc.). Then place the piece of metal on the wet spot with its respective cation. The top side of the metal should be kept dry. Then add several drops of 1 M $NaNO_3$ to the dotted lines drawn between each circle and the center of the filter paper. Be sure there is a continuous trail of $NaNO_3$ between each circle and the center. You may have to periodically dampen the filter paper with $NaNO_3$ during the experiment. **CAUTION:** *Handle these solutions with care. Some are poisonous, and some cause hard-to-remove stains. If a spill occurs, ask your teacher how to clean it up safely.*

DATA COLLECTION

8. Use metal M_1 (the one that is obviously copper) as the reference metal. Determine the potential of four cells by connecting M_1 to M_2, M_1 to M_3, M_1 to M_4, and M_1 to M_5. This is done by bringing the (+) lead in contact with one metal and the (−) lead in contact with the other. If the voltage displayed on the main screen of the calculator is (−), then reverse the leads. Wait about five seconds to take a voltage reading, and record the (+) value appearing on the calculator screen in Table 1 (round to the nearest 0.01 V). Also record which metal is the (+) terminal and which is (−), when the voltage value is positive. Use the same procedure and measure the potential of the other three cells, continuing to use M_1 as the reference electrode.

DATA TABLE 1

Voltaic cell (metals used)	Measured potential (V)	Metal number of (+) lead	Metal number of (−) lead
M_1/M_2			
M_1/M_3			
M_1/M_4			
M_1/M_5			

9. Go to Step 1 of Processing the Data. Use the method described in Step 1 to rank the five metals from the lowest (−) reduction potential to the highest (+) reduction potential. Then *predict* the potentials for the remaining six cell combinations.

10. Now return to your work station and *measure* the potential of the six remaining half-cell combinations. If the $NaNO_3$ salt bridge solution has dried, you may have to re-moisten it. Record each measured potential in Table 2.

DATA TABLE 2

	Predicted potential (V)	Measured potential (V)	Percentage error (%)
M_2/M_3			
M_2/M_4			
M_2/M_5			
M_3/M_4			
M_3/M_5			
M_4/M_5			

Name _____ Class _____ Date _____

Micro-Voltaic Cells *continued*

11. When you are finished, select QUIT and exit the DATAMATE program.

12. When you have finished, use forceps to remove each of the pieces of metal from the filter paper. Rinse each piece of metal with tap water. Dry it, and return it to the correct container. Remove the filter paper from the glass plate using the forceps, and discard it as directed by your teacher. Rinse the glass plate with tap water, making sure that your hands do not come in contact with wet spots on the glass.

PROCESSING THE DATA

1. After finishing Step 8 in the procedure, arrange the five metals (including M_1) in Data Table 2 from the lowest reduction potential at the top (most negative) to the highest reduction potential at the bottom (most positive). Metal M_1, the standard reference, will be given an arbitrary value of 0.00 V. If the other metal was correctly connected to the *negative* terminal, it will be placed *above* M_1 in the chart (with a negative E° value). If it was connected to the positive terminal, it will be placed below M_1 in the chart (with a positive E° value). The numerical value of the potential relative to M_1 will simply be the value that you measured. Record your results in Table 3.

DATA TABLE 3

Metal (Mₓ)	Lowest (−) reduction potential, E° (V)
	Highest (+) reduction potential, E° (V)

Then calculate the *predicted* potential of each of the remaining cell combinations shown in Table 2, using the reduction potentials you just determined (in Table 3). Record the predicted cell potentials in Table 2. Return to Step 10 in the procedure, and finish the experiment.

2. Calculate the percentage error for each of the potentials you measured in Step 10 of the procedure. Do this by comparing the measured cell potentials with the predicted cell potentials in Table 2.

Name _____ Class _____ Date _____

Micro-Voltaic Cells continued

Analysis

1. Examining Data Which metal had the highest reduction potential? Which had the lowest reduction potential? <u>**Metal M_4 (Ag/Ag^+) had the highest reduction**</u> <u>**potential. Metal M_2 (Zn/Zn^{2+}) had the lowest reduction potentials.**</u>

2. Examining Data Which combination of metals had the largest measured cell potential? <u>**The combination of metals M_2/M_4 (Zn/Ag) yielded a measured**</u> <u>**potential of 1.49 V.**</u>

3. Analyzing Results In Step 8, the metal M_1 was arbitrarily given a reduction potential of 0.00 V and used as the reference metal in determining the potential of each voltaic cell. The metal M_1 can be positively identified as being copper. According to the Table of Standard Reduction Potentials, Cu has a reduction potential of +0.34 V. Adjust all of the reduction potentials in Table 3 by adding +0.34 V. Use the Table of Standard Reduction Potentials found in your textbook to properly identify each of the unknown metals in Table 3.

Conclusions

1. Drawing Conclusions If metal M_4 were actually gold, would your measured cell potential have been higher or lower? Explain why. <u>**The substitution of**</u> <u>**gold for silver would have increased the cell potential for all combinations in**</u> <u>**this experiment. The metal gold (Au) has a reduction potential of +1.50 V,**</u> <u>**which is higher than silver (Ag) with a reduction potential of +0.80 V.**</u>

2. Drawing Conclusions If the battery back in the plane requires a 6 V power source, how can you arrange the electrochemical cell to yield the necessary voltage? <u>**An electrochemical cell using Ag and Zn should yield a standard**</u> <u>**potential of +1.56 V. If four cells were connected in series, the result would**</u> <u>**be a voltage of +6.24 V.**</u>

Air Pressure and Piston Design

Teacher Notes

TIME REQUIRED One 50-minute lab period

LAB RATINGS   Easy ◄—¹——²——³——⁴—► Hard

Teacher Preparation–2
Student Setup–2
Concept Level–2
Cleanup–1

SKILLS ACQUIRED

Collecting data
Experimenting
Organizing and analyzing data
Interpreting
Identifying/recognizing patterns

SCIENTIFIC METHODS

Make Observations Students will collect pressure and volume data using a calculator-interfaced pressure sensor.

Analyze the Results In Analysis questions 1–6, students will analyze the data from their experiments and identify patterns.

Draw Conclusions In Conclusions questions 1–5, students will interpret the data and relate their results to real-world examples.

MATERIALS (PER LAB GROUP)

- LabPro or CBL2 interface
- syringe, plastic, 20 mL
- TI graphing calculator

- Vernier gas pressure sensor or Vernier pressure sensor

Graphing Calculator and Sensors

TIPS AND TRICKS

Students should have the DataMate program loaded on their graphing calculators. Refer to Appendix B of Vernier's *Chemistry with Calculators* for instructions.

Not all models of TI graphing calculators have the same amount of memory. If possible, instruct students to clear all calculator memory before loading the DataMate program.

This experiment has been written to use either the original pressure sensor or the newer gas pressure sensor. The gas pressure sensor is equipped with auto-ID capability, and a calibration in units of kPa (kilopascals) will automatically be loaded when the DataMate program is started. You may select other pressure units (mm Hg or atm) when using DataMate by selecting the Setup menu.

It is not necessary to perform a new calibration when using the gas pressure sensor or pressure sensor. Simply use the calibration file that is stored in the DataMate program.

NOTES ON TECHNIQUE

Before they view graphs on the calculator, remind students how to use the arrow keys to trace the data points on the graph.

The procedure found in the Extensions section of this experiment requires students to exit the DataMate program and manually clear their calculator lists. You may wish to review this procedure with your students before beginning this experiment.

Experimental Setup
TIPS AND TRICKS

To save time, you may prefer to do Step 1 of the student procedure before the start of class.

One potential source of error in this experiment is the small inside volume of the white stem that leads to the inside of the gas pressure sensor. The volume of this space is about 0.8 mL. This means that when students enter a volume of 10 mL (as read on the syringe), the volume is really about 10.8 mL. To compensate for this error, you can have your students add 0.8 mL to each of the volumes they enter. They will get better results for the value of the exponent, b, in Step 7.

Answers

CONCLUSIONS

2. The correct formula to choose for an inverse relationship is $k = PV$. The sample data table lists the calculated k values for this experiment. The average k value is 1028 mL • kPa.

DATA TABLES WITH SAMPLE DATA

Volume (mL)	Pressure (kPa)	Constant, k (P/V or P•V)
5.0	204.6	1023
7.5	136.8	1026
10.0	103.3	1033
12.5	82.1	1026
15.0	69.9	1048
17.5	58.8	1028
20.0	50.7	1013

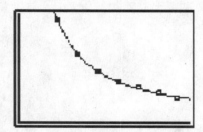

Pressure Versus Volume Graph with Regression Curve

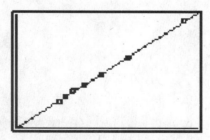

Pressure Versus 1/Volume Graph with Regression Line

Chemical Engineering Lab

PROBEWARE LAB

Air Pressure and Piston Design

Recently, your design firm has been contracted to design the piston and cylinder for an air compressor. As the staff mechanical engineer, you are responsible for this task. You have been issued a list of specifications for the compressor. The cylinder must have a total volume of 2 L and a compressed volume of 500 mL. Before you can determine the best material to use for the cylinder and piston, you must calculate how much pressure the walls of the cylinder must withstand.

To better understand the relationship between pressure and volume of a confined gas, you will need to investigate Boyle's Law. In 1662, Robert Boyle established that there was a mathematical relationship between the pressure of a confined gas and its volume when the temperature and amount of gas remained constant.

In this experiment, your team will determine the relationship between the pressure and volume of a confined gas. The gas you use will be air, and it will be confined in a syringe connected to a pressure sensor. As you move the piston of the syringe, the volume and pressure of the gas contained within will change. The pressure change will be monitored using a pressure sensor. It is assumed that the temperature and amount of gas will remain constant throughout the experiment. Pressure and volume data pairs will be collected and analyzed in this experiment. Using your collected data, you will determine the mathematical relationship between the pressure and volume of the confined gas. Then, you will be able to apply what you have learned to the piston design problem.

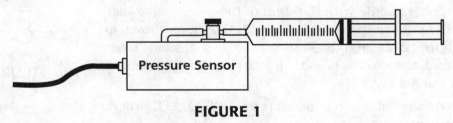

Pressure Sensor

FIGURE 1

OBJECTIVES

Measure pressure changes.

Graph pressure-volume data pairs.

Calculate pressure and volume relationship.

Relate Boyle's law to real-world applications.

EQUIPMENT

- LabPro or CBL2 interface
- syringe, plastic, 20 mL
- TI graphing calculator
- Vernier gas pressure sensor or Vernier pressure sensor

Air Pressure and Piston Design *continued*

SAFETY

Although the equipment and procedures used in this experiment are not particularly hazardous, always exercise caution when in labs, because there may be other hazards present in the lab room.

Procedure

EQUIPMENT PREPARATION

1. Prepare the pressure sensor and an air sample for data collection.

   a. Plug the pressure sensor into Channel 1 of the LabPro or CBL2 interface. Use the link cable to connect the TI graphing calculator to the interface. Firmly press in the cable ends.

   b. With the 20 mL syringe disconnected from the pressure sensor, move the piston of the syringe until the front edge of the inside black ring is positioned at the 10.0 mL mark.

   c. Attach the 20 mL syringe to the valve of the pressure sensor.

   • Newer Vernier gas pressure sensors have a white stem protruding from the end of the sensor box—attach the syringe directly to the white stem with a gentle half turn.

   • Older Vernier pressure sensors have a 3-way valve at the end of a plastic tube leading from the sensor box. Before attaching the 20 mL syringe, align the blue handle with the stem of the 3-way valve that will *not* have the syringe connected to it, as shown in **Figure 2** at the right—this will close this stem. Then attach the syringe directly to the remaining open stem of the 3-way valve.

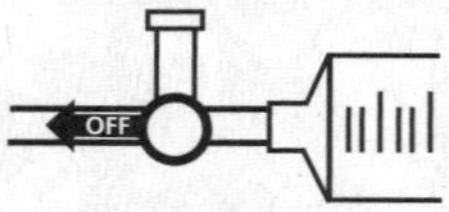

   FIGURE 2

2. Turn on the calculator, and start the DATAMATE program. Press CLEAR to reset the program.

3. Set up the calculator and interface for a gas pressure sensor or pressure sensor.

   a. Select SETUP from the main screen.

   b. If the calculator displays a pressure sensor set to kPa in CH 1, proceed directly to Step 4. If it does not, continue with this step to set up your sensor manually.

   c. Press ENTER to select CH 1.

   d. Select PRESSURE from the SELECT SENSOR menu.

   e. Select the correct pressure sensor (GAS PRESSURE SENSOR or PRESSURE SENSOR) from the PRESSURE menu.

   f. Select the calibration listing for units of KPA.

4. Set up the data-collection mode.

　a. To select MODE, press ▲ once and press [ENTER].

　b. Select EVENTS WITH ENTRY from the SELECT MODE menu.

　c. Select OK to return to the main screen.

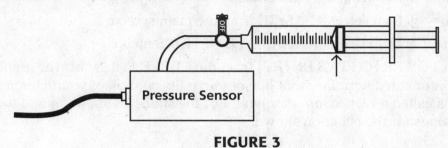

FIGURE 3

DATA COLLECTION

5. You are now ready to collect pressure and volume data. It is best for one person to take care of the gas syringe and for another to operate the calculator.

　a. Select START to begin data collection.

　b. Move the piston so the front edge of the inside black ring (see **Figure 3**) is positioned at the 5.0 mL line on the syringe. Hold the piston firmly in this position until the pressure value displayed on the calculator screen stabilizes.

　c. Press [ENTER], and type in "5," the gas volume (in mL), on the calculator. Press [ENTER] to store this pressure-volume data pair.

　d. To collect another data pair, move the syringe to 7.5 mL. When the pressure reading stabilizes, press [ENTER] and enter "7.5" as the volume.

　e. Continue with this procedure using volumes of 10.0, 12.5, 15.0, 17.5, and 20.0 mL.

　f. Press [STOP] when you have finished collecting data.

6. Examine the data pairs on the displayed graph. As you move the cursor right or left, the volume (X) and pressure (Y) values of each data point are displayed below the graph. Record the pressure (round to the nearest 0.1 kPa) and volume data values in your data table.

Volume (mL)	Pressure (kPa)	Constant, k (P/V or $P \cdot V$)

7. Based on the graph of pressure versus volume, decide what kind of mathematical relationship exists between these two variables. Is it a direct relationship (greater pressures lead to greater volumes) or an inverse relationship
(greater pressures lead to smaller volumes)? You can use the graphing calculator to see if you made the right choice:

a. Press [ENTER], then select ANALYZE from the main screen.

b. Select CURVE FIT from the ANALYZE OPTIONS menu.

c. Select POWER (CH 1 VS ENTRY) from the CURVE FIT menu. The graphing
calculator can determine what line or curve fits a graph of your data best
(this is called a *regression analysis*). For the statistics shown, *a* and *b* are
the values for the equation shown.

$$y = ax^b$$

or

$$(\text{pressure}) = a(\text{volume})^b$$

If the relationship is direct, *b* will have a positive value. If the relationship is
inverse, *b* will have a negative value.

d. To display the regression curve that best fits your graph of pressure versus
volume, press [ENTER]. If you have correctly determined the mathematical
relationship, the curve should very nearly fit the points on the graph (that
is, pass through or near the plotted points).

8. (optional) Print a graph of pressure versus volume, with a regression line
displayed.

Analysis

1. Examining Data When the syringe volume is moved from 5.0 mL to 10.0 mL,
how did your pressure values change? Show the pressure values in your answer.

When the syringe volume was doubled, the pressure was reduced by half. If

the pressure at 5.0 mL was 204.6 kPa, then the pressure at 10.0 mL would be

about 102.3 kPa.

2. Identifying/Recognizing Patterns Based on your data, if you take 1.0 L of air
and place it in a 2.0 L cylinder, how would the pressure change? _____

The pressure in the cylinder would be one-half of the atmospheric pressure.

Air Pressure and Piston Design *continued*

3. Identifying/Recognizing Patterns How would the pressure change if 1.0 L of air was placed in a 3.0 L cylinder? **The pressure in the cylinder would be one-third of the atmospheric pressure.**

4. Examining Data When the syringe volume is moved from 20.0 mL to 10.0 mL, what does your data show happens to the pressure? Show the pressure values in your answer. **When the syringe volume was halved, the pressure was doubled. If the pressure at 20.0 mL was 50.7 kPa, then the pressure at 10.0 mL would be about 101.4 kPa.**

5. Identifying/Recognizing Patterns How would the pressure change if 1.0 L of air was compressed in a cylinder to 500 mL? _____

The pressure in the cylinder would be twice the atmospheric pressure.

6. Describing Events What experimental factors are assumed to be constant in this experiment? **In this experiment, it is assumed that the temperature and number of molecules of gas sample are constant.**

Conclusions

1. Drawing Conclusions From your previous answers and the shape of the curve in the graph of pressure versus volume, is the relationship between the pressure and volume of a confined gas direct or inverse? Explain your answer.

Based on the data, the relationship appears to be inverse. The volume data appears to decrease proportionally as the pressure data increases. The shape of the pressure-volume plot is a simple inverse relationship.

2. Applying Conclusions Calculating the proportionality constant, k, is one way to determine if a relationship is inverse or direct. If this relationship is direct, $k = P/V$. If it is inverse, $k = PV$. Based on your answer to Conclusions question 1, choose one of these formulas and calculate k for the seven ordered pairs in your data table (divide or multiply the P and V values). Show the answers in the third column of the Data Collection table.

3. Evaluating Data How *constant* were the values for k you obtained in Conclusions question 2? Minor variation can be expected, but the values for k should be relatively constant. **The k values from this experiment should be**

relatively constant with only minor variations.

4. Interpreting Information Write an equation for Boyle's law using P, V, and k. Write a definition explaining Boyle's law. **The equation for Boyle's Law is**

$k = PV$. **The volume of a gas at constant temperature is inversely propor-**

tional to the pressure.

5. Applying Conclusions Calculate the gas pressure exerted on the cylinder walls when compressed at each of the following volumes. Assume that the initial atmospheric pressure is 100.0 kPa.

Compressed volume (L)	Gas volume (L)	Pressure (kPa)
1.00	1.0	100.0
0.50	1.0	**200.0**
0.25	1.0	**400.0**
2.00	1.0	**50.0**

Extensions

1. To confirm that an inverse relationship exists between pressure and volume, a graph of pressure versus *reciprocal of volume* (1/volume or volume^{-1}) may also be plotted. To do this using your calculator:

a. Press [ENTER], then return to the main screen.

b. Select QUIT to quit DATAMATE program. (Then press [ENTER] on a TI-83 Plus or TI-73).

c. Create a new data list, reciprocal of volume, based on your original volume data, following the steps shown below for your type of calculator.

TI-73 Calculators

d. To view the data lists, press [LIST].

e. Move the cursor up and to the right until the L3 heading is highlighted.

f. Create a list of *1/volume* values in L3. First press [2nd] [STAT], and select L1. Then press [2nd] [x^{-1}] [ENTER].

Air Pressure and Piston Design *continued*

TI-83 and TI-83 Plus Calculators

d. To view the data lists, press [STAT] to display the EDIT menu, and select Edit.

e. Move the cursor up and to the right until the L3 heading is highlighted.

f. Create a list of *1/volume* values in L3 by pressing [2nd] [L₁] [x⁻¹] [ENTER].

TI-86 Calculators

d. To view the data lists, press [2nd] [STAT] and select EDIT.

e. Move the cursor up and to the right until the L3 heading is highlighted.

f. Create a list of *1/volume* values in L3 by pressing [NAMES] [L₁] [2nd] [x⁻¹] [ENTER].

g. Press [2nd] [QUIT] when you are finished with this step.

TI-89, TI-92, and TI-92 Plus Calculators

d. Press [APPS], then select Home.

e. On a TI-89 calculator, create a list of *1/volume* values in L3 by pressing [CLEAR] [1] [÷] [ALPHA] [L] [1] [STO▸] [ALPHA] [L] [3] [ENTER]. On a TI-92 or TI-92 Plus, press [CLEAR] [1] [÷] [L] [1] [STO▸] [L] [3] [ENTER].

2. Follow this procedure to determine what curve best fits your graph of pressure versus 1/volume:

  a. Restart the DATAMATE program.

  b. Select ANALYZE from the main screen.

  c. Select CURVE FIT from the ANALYZE OPTIONS menu.

  d. Select LINEAR (CH1 VS CH2). Note that CH1 is pressure and CH2 is 1/volume. The linear-regression statistics for these two lists are displayed for the equation in the form

$$y = ax + b$$

where x is 1/volume, y is pressure, a is a proportionality constant, and b is the y-intercept.

  e. To display the linear-regression curve on the graph of pressure versus 1/volume, press [ENTER]. If the relationship between P and V is an inverse relationship, the plot of P versus $1/V$ should be direct; that is, the curve should be linear and pass through (or near) the origin. Examine your graph to see if this is true for your data.

3. (optional) Print a copy of the graph of pressure versus 1/volume, with the linear regression curve displayed.

Evaporation and Ink Solvents

Teacher Notes

TIME REQUIRED One 50-minute lab period

LAB RATINGS

Easy ◄———¹——²——³——⁴——► Hard

Teacher Preparation–3
Student Setup–3
Concept Level–2
Cleanup–2

SKILLS ACQUIRED

Collecting data
Experimenting
Organizing and analyzing data
Interpreting
Drawing real-world conclusions

SCIENTIFIC METHODS

Form a Hypothesis In Procedure Steps 13 and 15, students will predict the change in temperature based on molecular weight and collected data.

Analyze the Results In Analysis questions 1–5, students will analyze the data and results from their experiment.

Draw Conclusions In Conclusions questions 1–4, students will interpret the data and apply them to the objectives of the experiment.

MATERIALS (PER LAB GROUP)

- 1-butanol
- 1-propanol
- ethanol (ethyl alcohol)
- filter paper, 2.5 cm × 2.5 cm (6 pieces)
- LabPro or CBL2 interface
- masking tape
- methanol (methyl alcohol)
- *n*-hexane
- *n*-pentane
- rubber bands, small (2)
- TI graphing calculator
- Vernier temperature probes (2)

SAFETY CAUTIONS

Because several of these liquids are highly volatile, keep the room well ventilated. Cap the test tubes or bottles at times when the experiment is not being performed. The experiment should not be performed near any open flames.

When using chemicals, students should wear aprons, gloves, and goggles.

Graphing Calculator and Sensors
TIPS AND TRICKS

Students should have the DataMate program loaded on their graphing calculators. Refer to Appendix B of Vernier's *Chemistry with Calculators* for instructions. The temperature calibrations that are stored in the DataMate data-collection program will work fine for this experiment. No calibration is necessary for the temperature probes.

Not all models of TI graphing calculators have the same amount of memory. If possible, instruct students to clear all calculator memory before loading the DataMate program.

Wrapped probes provide more-uniform liquid amounts and generally greater ΔT values than bare probes. Chromatography paper, filter paper, and various other paper types work well. Snug-fitting rubber bands can be made by cutting short sections from a small rubber hose. Surgical tubing works well. Orthodontist's rubber bands are also a good size.

The Vernier stainless steel temperature probe and CBL temperature probe will plug directly into CH1 on the Vernier LabPro or CBL2 interface. If you are using the Vernier direct-connect temperature probe, you will need a DIN-BTA (formerly CBL-DIN) adapter to convert from the 5-pin Din connector to the BTA connector.

NOTES ON TECHNIQUE

When viewing graphs on the calculator, students should use the arrow keys to trace the data points on the graph.

If students are using only a single temperature probe and would like to see two or three substances on the same graph, instruct them to store the first and/or second data set before beginning the next substance. From the Main Screen of DataMate, the Store Latest Run feature can be found under the Tools menu. The program will permit storing only up to two runs. If more than one sensor is used at a time, the Store Latest Run feature will not work.

Experimental Setup
TIPS AND TRICKS

Perform this experiment in a fume hood or in a well-ventilated classroom.

Other liquids can be substituted. Although it has a somewhat larger ΔT, 2-propanol can be substituted for 1-propanol. Some petroleum ethers have a high percentage of hexane and can be used in place of propanols. Other alkanes of relatively high purity, such as *n*-heptane or *n*-octane can also be used. Water, with a ΔT value of about 5°C, emphasizes the effect of hydrogen bonding on a low-molecular-weight liquid. However, students might have difficulty comparing its hydrogen bonding capability with that of the alcohols used.

Other properties besides ΔT values vary with molecular size and consequent size of intermolecular forces of attraction. Viscosity increases noticeably from methanol through 1-butanol. The boiling temperatures of methanol, ethanol, 1-propanol, and 1-butanol are 65°C, 78°C, 97°C, and 117°C, respectively.

Sets of the liquids can be supplied in 13 mm × 100 mm test tubes stationed in stable test-tube racks. This method uses very small amounts of the liquids. Alternatively, the liquids can be supplied in sets of small bottles kept for future use. Adjust the level of the liquid in each container so that the level will be above the top edge of the filter paper.

Answers

ANALYSIS

5. Graph of ΔT versus molecular weight:

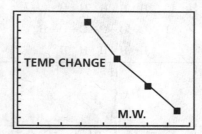

DATA TABLE WITH SAMPLE DATA

Substance	T_1 (°C)	T_2 (°C)	ΔT ($T_1 - T_2$) (°C)	Predicted ΔT(°C)	Explanation
Ethanol	23.5	15.2	8.3		
1-propanol	23.0	18.1	4.9		
1-butanol	23.2	21.5	1.7	varies (< 4.9°C)	higher molecular weight than 1-propanol (both have H-bonds)
n-pentane	23.0	6.9	16.1	varies (> 8.3°C)	higher molecular weight than either, but no H-bonding
Methanol	22.9	9.8	13.1	varies (> 8.3°C)	lower molecular weight than ethanol (both have H-bonds)
n-hexane	23.2	11.2	12.0	varies (< 16.1°C)	higher molecular weight than n-pentane; also no H-bonding

PRE-LAB RESULTS

Substance	Formula	Structural formulas	Molecular weight (amu)	Hydrogen bond (yes or no)
Ethanol	C_2H_5OH	H—C—C—O—H	46	yes
1-propanol	C_3H_7OH	H—C—C—C—O—H	60	yes
1-butanol	C_4H_9OH	H—C—C—C—C—O—H	74	yes
n-pentane	C_5H_{12}	H—C—C—C—C—C—H	72	no
Methanol	CH_3OH	H—C—O—H	32	yes
n-hexane	C_6H_{14}	H—C—C—C—C—C—C—H	86	no

Chemical Engineering Lab   **PROBEWARE LAB**

Evaporation and Ink Solvents

You are an organic chemist working for a company that manufactures various types of ink. You have been asked to create a calligrapher's ink that dries quickly at room temperature. The company feels that such a product would be a big hit because faster-drying ink would cause less distortion to the paper on which it is used. Ink consists of two components: a pigment, or coloring agent, and a solvent. The pigment is what gives the ink its color. The solvent is the chemical in which the pigment is dissolved. Your job is to select a solvent that will evaporate quickly.

To determine the best solvent to use, you will test two types of organic compounds—alkanes and alcohols. To establish how and why these substances evaporate, you will test four alcohols and two alkanes. From your results, you will be able to predict how other alcohols and alkanes will evaporate.

The two alkanes you will test are pentane, C_5H_{12}, and hexane, C_6H_{14}. Alkanes contain only carbon and hydrogen atoms, whereas alcohols also contain the –OH functional group. In this experiment, two of the alcohols you will test are methanol, CH_3OH, and ethanol, C_2H_5OH. To better understand why these substances evaporate, you will examine the molecular structure of each for the presence and relative strength of hydrogen bonding and London dispersion forces.

The process of evaporation requires energy to overcome the intermolecular forces of attraction. For example, when you perspire on a hot day, the water molecules in your perspiration absorb heat from your body and evaporate. The result is a lowering of your skin temperature known as evaporative cooling.

In this experiment, temperature probes will be placed into small containers of your test substances. When the probes are removed, the liquid on the temperature probes will evaporate. The temperature probes will monitor the temperature change. Using your data, you will determine the temperature change, ΔT, for each substance and relate that information to the substance's molecular structure and presence of intermolecular forces.

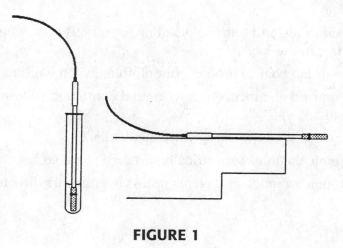

FIGURE 1

Evaporation and Ink Solvents *continued*

OBJECTIVES

Measure temperature changes.

Calculate changes in temperature.

Relate temperature changes to molecular bonding.

Predict temperature changes for various liquids.

MATERIALS

- 1-butanol
- 1-propanol
- ethanol (ethyl alcohol)
- filter paper pieces, 2.5 cm × 2.5 cm (6 pieces)
- masking tape
- methanol (methyl alcohol)
- *n*-hexane
- *n*-pentane
- rubber bands, small (2)

EQUIPMENT

- LabPro or CBL2 interface
- TI graphing calculator
- Vernier temperature probes (2)

SAFETY

- Wear safety goggles when working around chemicals, acids, bases, flames, or heating devices. Contents under pressure may become projectiles and cause serious injury.

- If any substance gets in your eyes, notify your instructor immediately and flush your eyes with running water for at least 15 minutes.

- Use flammable liquids only in small amounts.

- When working with flammable liquids, be sure that no one else in the lab is using a lit Bunsen burner or plans to use one. Make sure there are no other heat sources present.

- Secure loose clothing, and remove dangling jewelry. Do not wear open-toed shoes or sandals in the lab.

- Wear an apron or lab coat to protect your clothing when working with chemicals.

- Never return unused chemicals to the original container; follow instructions for proper disposal.

- Always use caution when working with chemicals.

- Never mix chemicals unless specifically directed to do so.

- Never taste, touch, or smell chemicals unless specifically directed to do so.

Name _____ Class _____ Date _____

Evaporation and Ink Solvents *continued*

Pre-Lab Procedure

Before doing the experiment, complete the pre-lab table below. The name and formula are given for each compound. Draw a structural formula for a molecule of each compound. Then determine the molecular weight of each of the molecules. Dispersion forces exist between any two molecules and generally increase as the molecular weight of the molecule increases. Next, examine each molecule for the presence of hydrogen bonding. Before hydrogen bonding can occur, a hydrogen atom must be bonded directly to an N, O, or F atom within the molecule. Record whether each molecule has hydrogen-bonding capability.

Substance	Formula	Structural formulas	Molecular weight	Hydrogen bond (yes or no)
Ethanol	C_2H_5OH			
1-propanol	C_3H_7OH			
1-butanol	C_4H_9OH			
n-pentane	C_5H_{12}			
Methanol	CH_3OH			
n-hexane	C_6H_{14}			

Procedure
EQUIPMENT PREPARATION

1. Obtain and wear goggles! **CAUTION:** *The compounds used in this experiment are flammable and poisonous. Avoid inhaling their vapors. Avoid their contact with your skin or clothing. Be sure there are no open flames in the lab during this experiment. Notify your teacher immediately if an accident occurs.*

2. Plug temperature probe 1 into Channel 1 and temperature probe 2 into Channel 2 of the LabPro or CBL2 interface. Use the link cable to connect the TI graphing calculator to the interface. Firmly press in the cable ends.

3. Turn on the calculator, and start the DATAMATE program. Press [CLEAR] to reset the program.

4. Set up the calculator and interface for two temperature probes.

    a. Select SETUP from the main screen.

    b. If the calculator displays two temperature probes, one in CH 1 and another in CH 2, proceed directly to Step 5. If it does not, continue with this step to set up your sensor manually.

    c. Press [ENTER] to select CH 1.

    d. Select TEMPERATURE from the SELECT SENSOR menu.

    e. Select the temperature probe you are using (in degrees Celsius) from the TEMPERATURE menu.

f. Press ▼ once, then press ⌜ENTER⌝ to select CH2.

g. Select TEMPERATURE from the SELECT SENSOR menu.

h. Select the temperature probe you are using (in degrees Celsius) from the TEMPERATURE menu.

5. Set up the data-collection mode.

a. To select MODE, use ▲ to move the cursor to MODE and press ⌜ENTER⌝.

b. Select TIME GRAPH from the SELECT MODE menu.

c. Select CHANGE TIME SETTINGS from the TIME GRAPH SETTINGS menu.

d. Enter "3" as the time between samples in seconds.

e. Enter "80" as the number of samples. (The length of the data collection will be four minutes.)

f. Select OK to return to the setup screen.

g. Select OK again to return to the main screen.

6. Wrap probe 1 and probe 2 with square pieces of filter paper secured by small rubber bands as shown in **Figure 1.** Roll the filter paper around the probe tip in the shape of a cylinder. Hint: First slip the rubber band up on the probe, wrap the paper around the probe, and then finally slip the rubber band over the wrapped paper. The paper should be even with the probe end.

7. Stand probe 1 in the ethanol container and probe 2 in the 1-propanol container. Make sure the containers do not tip over.

8. Prepare two pieces of masking tape, each about 10 cm long, to be used to tape the probes in position during Step 9.

DATA COLLECTION

9. After the probes have been in the liquids for at least 30 seconds, select START to begin collecting temperature data. A live graph of temperature versus time for both probe 1 and probe 2 is being plotted on the calculator screen. The live readings are displayed in the upper-right corner of the graph, the reading for probe 1 first, the reading for probe 2 below it. Monitor the temperature for 15 seconds to establish the initial temperature of each liquid. Then simultaneously remove the probes from the liquids, and tape them so that the probe tips extend 5 cm over the edge of the table top as shown in **Figure 1.**

10. Data collection will stop after four minutes (or press the ⌜STO▶⌝ key to stop *before* four minutes have elapsed). On the displayed graph of temperature versus time, each point for probe 1 is plotted with a dot and each point for probe 2 with a box. As you move the cursor right or left, the time (X) and temperature (Y) values of each probe 1 data point are displayed below the graph. Based on your data, determine the maximum temperature, T_1, and minimum temperature, T_2. Record T_1 and T_2 for probe 1.

Press ▼ to switch the cursor to the curve of temperature versus time for probe 2. Examine the data points along the curve. Record T_1 and T_2 for probe 2.

11. For each liquid, subtract the minimum temperature from the maximum temperature to determine ΔT, the temperature change during evaporation.

12. Roll the rubber band up the probe shaft, and dispose of the filter paper as directed by your instructor.

13. Based on the ΔT values you obtained for these two substances, plus information in the pre-lab exercise, *predict* the ΔT value for 1-butanol. Compare its hydrogen-bonding capability and molecular weight with those of ethanol and 1-propanol. Record your predicted ΔT, and then explain how you arrived at this answer in the space provided. Do the same for *n*-pentane. It is not important that you predict the exact ΔT value; simply estimate a logical value that is higher, lower, or between the previous ΔT values.

14. Press ENTER to return to the main screen. Test your prediction in Step 13 by repeating Steps 6–12 using 1-butanol with probe 1 and *n*-pentane with probe 2.

15. Based on the ΔT values you have obtained for all four substances, plus information in the pre-lab exercise, predict the ΔT values for methanol and *n*-hexane. Compare the hydrogen-bonding capability and molecular weight of methanol and *n*-hexane with those of the previous four liquids. Record your predicted ΔT, and then explain how you arrived at this answer in the space provided.

16. Press ENTER to return to the main screen. Test your prediction in Step 15 by repeating Steps 6–12, using methanol with probe 1 and *n*-hexane with probe 2.

DATA TABLE

Substance	T_1 (°C)	T_2 (°C)	$\Delta T (T_1 - T_2)$ (°C)	Predicted ΔT(°C)	Explanation
Ethanol					
1-propanol					
1-butanol					
n-pentane					
Methanol					
n-hexane					

Analysis

1. **Analyzing Data** Which of the tested alcohols evaporated the fastest? Which alcohol had the largest ΔT value? What was the alcohol's molecular weight?

   Of the alcohols, methanol evaporated the fastest and had the largest

   ΔT value. The molecular weight of methanol is 32.

   _____

2. **Analyzing Data** Which of the alcohols tested evaporated the slowest? Which alcohol had the smallest ΔT value and molecular weight? _____

   Of the alcohols, 1-butanol evaporated the slowest and had the smallest

   ΔT value. The molecular weight of 1-butanol is 74.

   _____

3. **Analyzing Results** The alcohol 1-butanol and alkane *n*-pentane have similar molecular weights, but their tests resulted in very different ΔT values. Based on the information in your pre-lab data table, explain the difference in the ΔT values of these substances. **The substances *n*-pentane and 1-butanol have**

   similar molecular weights of 72 and 74. In addition to dispersion forces,

   1-butanol possesses hydrogen bonding between its molecules. This results in

   a stronger attraction and a slower rate of evaporation, which results in a

   smaller ΔT value.

4. **Analyzing Information** What types of intermolecular forces are evident in this experiment? **Dispersion forces and hydrogen bonding**

   _____

   _____

5. **Analyzing Data** Make a graph of your data with the molecular weight of each substance on the *x*-axis and ΔT on the *y*-axis.

   See page 189c for a sample graph.

Evaporation and Ink Solvents *continued*

Conclusions

1. Evaluating Results Alcohols with more than one –OH group are known as glycols. The substance ethylene glycol, CH_2OHCH_2OH, has a molecular weight of 62. Based on your data, would you expect it to have a larger or smaller ΔT than 1-propanol? Explain your answer using the results of this experiment.

Ethylene glycol has two –OH groups, whereas 1-propanol has only one. This

additional –OH group increases the amount of hydrogen bonding resulting in

a significantly smaller ΔT than 1-propanol.

_____

_____

2. Inferring Conclusions Of the substances tested in this experiment, which would work best as the solvent for the ink your company is developing?

The best ink solvent, based on rate of evaporation, would be *n*-pentane,

which had a ΔT of 16.1°C.

Extensions

1. Applying Results Using methanol and ethanol as solvent choices, test different types of mediums. Use the different medium choices in place of the filter paper to determine if the medium has any effect on the rate of evaporation.

A Leaky Reaction

Teacher Notes

TIME REQUIRED One 50-minute lab period

LAB RATINGS Easy ←——1——2——3——4——→ Hard

Teacher Preparation–3
Student Setup–2
Concept Level–4
Cleanup–2

SKILLS ACQUIRED

Collecting data
Experimenting
Organizing and analyzing data
Interpreting
Drawing real-world conclusions

SCIENTIFIC METHODS

Make Observations Students will collect absorbance data using a calculator-interfaced colorimeter.

Analyze the Results In Analysis questions 1–5, students will analyze the data from their experiments and calculate answers to a real-world example.

Draw Conclusions In Conclusions question 1, students will evaluate the methods used in this lab activity.

MATERIALS (PER LAB GROUP)

- beaker, 100 mL
- cuvette, plastic
- crstal violet solution, 2.5×10^{-5} M, 10 mL
- LabPro or CBL2 interface
- NaOH, 0.10 M, 10 mL
- stirring rod
- TI graphing calculator
- Vernier colorimeter
- water, distilled

Prepare the NaOH solution by adding enough water to 4.0 g of solid NaOH to bring the total volume to 1 L. Prepare the crystal violet solution by adding 0.020 g of crystal violet per 2 L of solution. (Note: If a milligram balance is unavailable, prepare a 2.5×10^{-4} M crystal violet solution by using 0.20 g per 2 L of solution. Then dilute 100 mL of this solution to a total volume of 1 L.) Crystal violet leaves stains if spilled—be careful when preparing the solution. It also stains glassware if left for an extended time in glass containers. A solution of 1 M HCl will remove stains from glassware.

SAFETY CAUTIONS

Sodium hydroxide is a corrosive solid and is known to cause skin burns upon contact. Wear gloves when working with this substance. When it is added to water, much heat is evolved. It is very dangerous to eyes, so wear face and eye protection when using this substance.

When using chemicals, students should wear aprons, gloves, and goggles.

Graphing Calculator and Sensors

TIPS AND TRICKS

Students should have the DataMate program loaded on their graphing calculators. Refer to Appendix B of Vernier's *Chemistry with Calculators* for instructions.

Temperature changes for the sample placed in the colorimeter should be very minor during the three-minute data collection period.

If enough time is available, have students perform a second trial of data collection.

NOTES ON TECHNIQUE

Instruct students how to enter data by hand into the lists within their calculators. If students complete the Extension question at the end of this activity, they will be required to work with the data lists outside of the DATAMATE calculator program.

Answers

ANALYSIS

1. See sample data table.

2. See sample data table.

3. Answers may vary.

4. Answers may vary.
   Sample answer: The absorbance of the exiting solution had completed two half-lives.

   Initial absorbance = 0.324

   $\frac{1}{2}$ initial absorbance = 0.162 (first half-life)

   $\frac{1}{4}$ initial absorbance = 0.081 (second half-life)

   Calculated half-life from question #3 = 138 s

   Reaction length = 2 (# of half-lives) $\times$ 138 s = 276 s

5. Sample answer: Distance the reacting crystal violet traveled:

   Flow velocity of solution = 2.0 m/s

   Reaction length = 276 s

   Distance traveled = 276 s $\times$ 2.0 m/s = 552 m

CONCLUSIONS

1. Improving accuracy:

- Apply a mathematical regression to the data, allowing an interpolation between discreet data points. This would improve the accuracy of the half-life calculation.

- Maintain a constant temperature during the reaction.

Sources of error:

- In the discharge channel scenario, the change in absorbance of the crystal violet solution upon mixing with the sodium hydroxide is not taken into account. The initial absorbance of the solution is measured before the two solutions mix together.

- It is assumed in this experiment that the crystal violet and sodium hydroxide are sufficiently mixed together to create a homogenous solution.

- The flow velocity may not be constant the entire length of the channel.

EXTENSIONS

1. Of the three graphs plotted, the graph of ln absorbance versus time is the only one that is linear. This identifies the reaction as being first order with respect to crystal violet concentration.

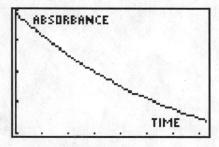

A versus time: reaction is not zero order

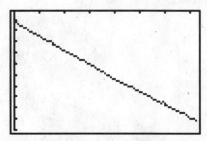

ln A versus time:
reaction is first order

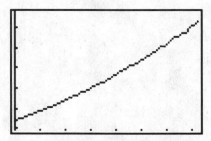

1/A versus time:
reaction is not second order

A Leaky Reaction

DATA TABLE AND SAMPLE DATA

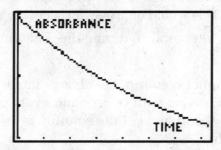

Graph of absorbance versus time for the crystal violet reaction

	Trial 1	Trial 2
Initial absorbance	0.312	0.288
Time at $\frac{1}{2}$ initial absorbance	138 s	141 s
Half-life of reaction	0 − 138 = 138 s	0 − 141 = 141 s
Time at $\frac{1}{4}$ initial absorbance	281 s	284 s
Half-life of reaction	281 − 138 = 143 s	284 − 141 = 143 s

Chemical Engineering Lab

A Leaky Reaction

A recent test of discharge water at the Hassenfrass paper mill indicates that the water has become contaminated with sodium hydroxide. It is believed that a buried pipe carrying a solution of sodium hydroxide has sprung a leak and the solution is leaking into the discharge water. Unfortunately, there are many pipes running near the discharge channels and it is difficult to easily identify which pipe is leaking. The leak could be anywhere along a 200 m section of the discharge channel. The pH of the discharge water has been measured to determine the molarity and volume of solution escaping into the discharge water.

The environmental analysis firm you work for has been called out to locate the leak. Field technicians have put together a plan, which would involve replacing the discharge water with an aqueous solution of the indicator crystal violet. As the crystal violet solution flows through the discharge channel, it will react with the sodium hydroxide. This reaction results in a change in the color of the crystal violet solution from dark violet to colorless. The field technicians have contacted you because they need to know specifics concerning the crystal violet reaction. Your job is to determine the reaction rate and half-life when crystal violet and sodium hydroxide react. The field technicians believe that they can use colorimetric analysis to determine how long the crystal violet exiting the channel has reacted with the sodium hydroxide. This information, along with the flow rate of the channel, can be used to calculate how far up the channel the leak is occurring.

During the reaction, the solution will change from a violet color to colorless. Using a colorimeter set to the green (565 nm) wavelength, you will monitor the absorbance of the crystal violet solution over time. It is assumed that the measured absorbance is proportional to the concentration of crystal violet.

OBJECTIVES

Measure absorbance values of a crystal violet solution.

Gather absorbance vs. time data.

Graph absorbance-time data pairs.

Estimate the half-life of a reaction.

Calculate reaction time.

MATERIALS
- crystal violet solution, 2.5×10^{-5} M
- NaOH, 0.10 M
- water, distilled

EQUIPMENT

- beaker, 100 mL
- cuvette, plastic
- LabPro or CBL2 interface
- stirring rod
- TI graphing calculator
- Vernier colorimeter

SAFETY

- Wear safety goggles when working around chemicals, acids, bases, flames, or heating devices. Contents under pressure may become projectiles and cause serious injury.

- If any substance gets in your eyes, notify your instructor immediately, and flush your eyes with running water for at least 15 minutes.

- Secure loose clothing, and remove dangling jewelry. Do not wear open-toed shoes or sandals in the lab.

- Wear an apron or lab coat to protect your clothing when working with chemicals.

- Never return unused chemicals to the original container; follow instructions for proper disposal.

- Always use caution when working with chemicals.

- Never mix chemicals unless specifically directed to do so.

- Never taste, touch, or smell chemicals unless specifically directed to do so.

Procedure
EQUIPMENT PREPARATION

1. Obtain and wear goggles.

2. Use a 10 mL graduated cylinder to obtain 10.0 mL of 0.10 M NaOH solution. **CAUTION:** *Sodium hydroxide solution is caustic. Avoid spilling it on your skin or clothing.* Use another 10 mL graduated cylinder to obtain 10.0 mL of 2.5×10^{-5} M crystal violet solution. **CAUTION:** *Crystal violet is a biological stain. Avoid spilling it on your skin or clothing.*

3. Plug the colorimeter into Channel 1 of the LabPro or CBL2 interface. Use the link cable to connect the TI graphing calculator to the interface. Firmly press in the cable ends.

4. Prepare a *blank* by filling an empty cuvette three-fourths full with distilled water. Seal the cuvette with a lid. To correctly use a colorimeter cuvette, remember the following:

   - All cuvettes should be wiped clean and dry on the outside with a tissue.

   - Handle cuvettes only by the top edge of the ribbed sides.

   - All solutions should be free of bubbles.

   - Always position the cuvette with its reference mark facing toward the white reference mark at the right of the cuvette slot on the colorimeter.

Name _____ Class _____ Date _____

A Leaky Reaction *continued*

5. Turn on the calculator, and start the DATAMATE program. Press [CLEAR] to reset the program.

6. Set up the calculator and interface for the colorimeter.

    a. Place the blank in the cuvette slot of the colorimeter, and close the lid.

    b. Select SETUP from the main screen.

    c. If the calculator displays COLORIMETER in CH 1, set the wavelength on the colorimeter to 565 nm. Then calibrate by pressing the AUTO CAL button on the colorimeter, and proceed directly to Step 7. If the calculator does not display COLORIMETER in CH 1, continue with this step to set up your sensor manually.

    d. Press [ENTER] to select CH 1.

    e. Select COLORIMETER from the SELECT SENSOR menu.

    f. Select CALIBRATE from the SETUP menu.

    g. Select CALIBRATE NOW from the CALIBRATION menu.

First Calibration Point

    h. Turn the wavelength knob of the colorimeter to the 0% T position. When the voltage reading stabilizes, press [ENTER]. Enter "0" as the percent transmittance.

Second Calibration Point

    i. Turn the wavelength knob of the colorimeter to the green LED position (565 nm). When the voltage reading stabilizes, press [ENTER]. Enter "100" as the percent transmittance.

    j. Select OK to return to the setup screen.

7. Set up the data-collection mode.

    a. To select MODE, press [▲] once and press [ENTER].

    b. Select TIME GRAPH from the SELECT MODE menu.

    c. Select CHANGE TIME SETTINGS from the TIME GRAPH SETTINGS menu.

    d. Enter "3" as the time between samples in seconds.

    e. Enter "60" as the number of samples. The length of the data collection will be three minutes.

    f. Select OK to return to the setup screen.

    g. Select OK again to return to the main screen.

DATA COLLECTION

8. You are now ready to begin monitoring data.

    a. To initiate the reaction, simultaneously pour the 10 mL portions of crystal violet and sodium hydroxide into a 100 mL beaker, and stir the reaction mixture with a stirring rod.

Name _____ Class _____ Date _____

A Leaky Reaction *continued*

    b. Empty the water from the cuvette. Rinse the cuvette with about 1 mL of the reaction mixture, and then fill it three-fourths full.

    c. Place the cuvette in the cuvette slot of the colorimeter, and close the lid.

    d. Monitor the absorbance reading on the main screen of the calculator for about 10 seconds (the absorbance reading should be gradually decreasing), then select START to begin data collection.

    e. During the three-minute data collection, observe the solution in the beaker as it continues to react.

    f. Data collection will end after three minutes.

    g. Discard the contents of the beaker and cuvette as directed by your teacher.

DATA TABLE

	Trial 1	Trial 2
Initial absorbance		
Time at $\frac{1}{2}$ initial absorbance		
Half-life of reaction		
Time at $\frac{1}{4}$ initial absorbance		
Half-life of reaction		

CLASS DATA

Lab Team	Half-life	Lab Team	Half-life
1		7	
2		8	
3		9	
4		10	
5		11	
6		12	

Analysis

1. Organizing Data A graph of absorbance versus time will be displayed. Use the ▶ or ◀ keys to examine the data points along the displayed curve of absorbance versus time. As you move the cursor right or left, the time (X) and absorbance (Y) values of each data point are displayed below the graph. Determine the initial absorbance of the crystal violet solution. Write the initial absorbance in your data table.

2. Examining Data Using the displayed graph, estimate the half-life of the reaction; move the cursor to a data point that is about half of the initial absorbance value. The *time* it takes the absorbance (or concentration) to be halved is known as the *half-life* for the reaction. Depending on your initial absorbance value, determine a second half-life for the reaction by moving the cursor to the data point that corresponds to a quarter of the initial absorbance value. Record the half-lives in your data table.

3. Analyzing Results Record the half-life of the crystal violet reaction found by each lab team in the class in the data table on the previous page. Using the

class data, calculate the average half-life. _____

4. Analyzing Results The field technicians filled the discharge channel with a 2.5×10^{-5} M solution of crystal violet. The initial absorbance of the solution was 0.488. The absorbance of the solution upon exiting the channel was 0.122. Using the average half-life calculated in Analysis question 3, calculate the length of time the crystal violet was reacting with the leaking sodium hydroxide.

_____

5. Analyzing Results It was determined that the solution flowed through the discharge channel at 2.0 m/s. Calculate the distance the reacting crystal violet

traveled. _____

_____

Conclusions

1. Evaluating Methods How might you increase the accuracy of the results

obtained in this lab exercise? Identify possible sources of error. _____

_____

_____

_____

Extensions

Designing Experiments Determine the reaction order with respect to crystal violet.

1. Analyze the data graphically to decide whether the reaction is zero, first, or second order with respect to crystal violet:

 • Zero order: If the current graph of absorbance versus time is linear, the reaction is *zero order*.

A Leaky Reaction *continued*

- First order: To see if the reaction is first order, it is necessary to plot a graph of the natural logarithm (ln) of absorbance versus time. If this plot is linear, the reaction is *first order*.

- Second order: To see if the reaction is second order, plot a graph of the reciprocal of absorbance versus time. If this plot is linear, the reaction is *second order*.

2. Follow these directions to create the natural log (ln) absorbance list in L_3 and the 1/absorbance list in L_4. To do this using your calculator, create a new data list based on your original volume data. Press [ENTER] to return to the main screen, and quit the DataMate program.

TI-73, TI-83, and TI-83 Plus Calculators

a. To view the lists, press [STAT] to display the EDIT menu and then select Edit.

b. To create a list of natural log (ln) of absorbance values in L_3, move the cursor up and to the right until the L_3 heading is highlighted, then press [LN] [2nd] [L2] [ENTER].

c. To create a list of reciprocal of absorbance values in L_4, move the cursor up and to the right until the L_4 heading is highlighted, then press [2nd] [L2] [x^{-1}] [ENTER].

d. Proceed to Step 3.

TI-86 Calculators

a. To view the lists, press [2nd] [STAT] and select EDIT.

b. To create a list of natural log (ln) of absorbance values in L_3, move the cursor up and to the right until the L_3 heading is highlighted, select NAMES, then press [L1] [2nd] [x^{-1}] [ENTER].

c. To create a list of reciprocal of absorbance values in L_4, move the cursor up and to the right until the L_4 heading is highlighted, then select L2, then press [2nd] [x^{-1}] [ENTER].

d. Press [2nd] [QUIT] when you are finished with this step, and then proceed to Step 3.

TI-89 Calculators

a. Press [APPS], then select Home.

b. To create a list of natural log (ln) of absorbance values in L_3, press [CLEAR] [2nd] [LN] [ALPHA] [L] [2] [ )] [STO▶] [ALPHA] [L] [3] [ENTER].

c. To create a list of reciprocal of absorbance values in L_4, press [CLEAR] [1] [÷] [ALPHA] [L] [2] [STO▶] [ALPHA] [L] [4] [ENTER]. Proceed to Step 3.

| A Leaky Reaction *continued*

TI-92 and TI-92 Plus Calculators

a. Press [APPS], then select Home.

b. To create a list of natural log (ln) of absorbance values in L3, press [CLEAR] [LN] [L] [2] [STO►] [L] [3] [ENTER].

c. To create a list of reciprocal of absorbance values in L4, press [CLEAR] [1] [÷] [L] [2] [STO►] [L] [4] [ENTER]. Proceed to Step 3.

3. Follow this procedure to calculate regression statistics and to plot a best-fit regression line on your graph of absorbance, ln absorbance, or reciprocal of absorbance versus time:

a. Restart the DATAMATE program.

b. Select ANALYZE from the main screen.

c. Select CURVE FIT from the ANALYZE OPTIONS menu.

d. Select LINEAR (CH 1 VS ENTRY), LINEAR (CH 2 VS ENTRY), or LINEAR (CH 3 VS ENTRY) from the CURVE FIT menu. Note that CH1 is absorbance, CH2 is natural log (ln) absorbance, and CH3 is reciprocal of absorbance. The linear-regression statistics for these two lists you select are displayed for the equation in the form

$$y = ax + b$$

where x is time; y is absorbance, ln absorbance, or reciprocal absorbance; a is the slope; and b is the y-intercept.

e. To display the linear-regression curve on the graph, press [ENTER]. Examine your graph to see if the relationship is linear.

f. (optional) Print a copy of the graph.

g. To view a graph of the two other lists, press [ENTER] to return to the ANALYZE OPTIONS menu, and repeat Steps 3c–3f.

Solubility and Chemical Fertilizers

Teacher Notes

TIME REQUIRED One 50-minute lab period (requires a full class period)

LAB RATINGS   Easy ◄——1——2——3——4——► Hard
    Teacher Preparation–2
    Student Setup–2
    Concept Level–2
    Cleanup–2

SKILLS ACQUIRED
    Collecting data
    Experimenting
    Organizing and analyzing data
    Interpreting
    Drawing real-world conclusions

SCIENTIFIC METHODS

Analyze the Results In Analysis questions 1–5, students will analyze the data and results from their experiment.

Draw Conclusions In Conclusion question 1, students will use the Internet to compare their solubility values with accepted solubilities.

MATERIALS (PER LAB GROUP)

- beaker, 250 mL
- beaker, 400 mL
- graduated cylinder, 10 mL
- hot plate
- LabPro or CBL2 interface
- potassium nitrate, KNO_3, 20 g
- ring stand

- stirring rod
- test-tube rack
- test tube, 20 mm × 150 mm (4)
- TI graphing calculator
- utility clamp (2)
- Vernier temperature probe (2)

SAFETY CAUTIONS

Potassium nitrate is a strong oxidant. It poses a fire and explosion risk when it is heated or in contact with organic material. It is also known to be a skin irritant.

When using hot plates, students should keep the temperature probe wire well away from the hot plate, as mentioned in Step 8 of the procedure.

When using chemicals, students should wear aprons, gloves, and goggles.

Graphing Calculator and Sensors
TIPS AND TRICKS

Students should have the DataMate program loaded on their graphing calculators. Refer to Appendix B of Vernier's *Chemistry with Calculators* for instructions. The temperature calibrations that are stored in the DataMate data-collection program will work fine for this experiment. No calibration is necessary for the temperature probes.

Not all models of TI graphing calculators have the same amount of memory. If possible, instruct students to clear all calculator memory before loading the DataMate program.

The Vernier stainless steel temperature probe and CBL temperature probe will plug directly into CH1 on the Vernier LabPro or CBL2 interface. If you are using the Vernier direct-connect temperature probe, you will need a DIN-BTA (formerly CBL-DIN) adapter to convert from the 5-pin Din connector to the BTA connector.

NOTES ON TECHNIQUE

When viewing graphs on the calculator, students should use the arrow keys to trace the data points on the graph.

Demonstrate to students how the calculator can be used to apply a best-fit line to data. The DataMate calculator program has built-in linear regression capabilities. The linear fit option can be found under the Analyze menu.

Once a best-fit line has been applied to the data on a graph, the calculator can be used to interpolate values between data points on the graph. Show students how they can use this feature to help answer the Analysis questions.

Experimental Setup
TIPS AND TRICKS

Students should be well prepared for this experiment in order to comfortably complete it during a 50-minute class period. Remind them of the tips in Step 11 of the student procedure. Starting hot-water baths promptly will save time. The 2 g and 4 g samples do not need to be heated above 60°C to have all of the solid dissolve; thus, the water bath does not need to be at 90°C for these two trials. Toward the end of the class period, some students may have to use cool-water baths, especially for the 2 g and 4 g samples.

Answers

ANALYSIS

1. See graph in Sample Data section below.

DATA TABLES WITH SAMPLE DATA

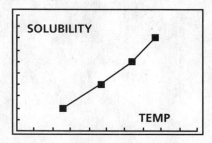

Solubility of KNO₃ (per 100 g H₂O) Versus Temperature

Trial	Solubility (g/100 g H₂O)	Temp (°C)
1	4.0×10^2	25.5
2	8.0×10^2	47.2
3	1.2×10^3	64.0
4	1.6×10^3	76.4

Name _____ Class _____ Date _____

Chemical Engineering Lab

Solubility and Chemical Fertilizers

You have been hired as a chemist for a fertilizer manufacturer. The manufacturer is building a facility to manufacture a new brand of fertilizer. One of the chief components of this new fertilizer is potassium nitrate (KNO_3). Large pipes carrying the potassium nitrate solution feed into enormous vats where all the ingredients are mixed together. Your job is to create a solubility table so that engineers will know at what temperature the solution must be kept. If the solution's temperature dips too low, potassium nitrate crystals may form and clog the pipes. Such an incident could stop production, costing the company a lot of money.

Solubility is the maximum amount of a solute that can dissolve in a solvent. When additional solute is added past a solution's solubility, the additional solute will not dissolve. Temperature can have a great effect on solubility. Consequently, solubility values always include a temperature. For example, the solubility of sodium nitrate at 20°C is 88.0 g per 100 g of H_2O. If the temperature of the solvent is raised to 50°C, the solubility increases to 114.0 g per 100 g of H_2O. The solubility of most solid solutes increases with temperature. Solubility tables are a very useful tool in the preparation of concentrated solutions.

In this experiment, you will completely dissolve different quantities of potassium nitrate, KNO_3, in the same volume of water at a high temperature. As each solution cools, you will monitor temperature using a temperature probe and observe the precise instant that solid crystals start to form. At this moment, the solution is saturated and contains the maximum amount of solute at that temperature. Thus, each data pair consists of a *solubility* value (g of solute per 100 g H_2O) and a corresponding *temperature*. A graph of the temperature-solubility data, known as a solubility curve, will be plotted using the TI calculator.

OBJECTIVES

Measure temperature values of salt solutions.

Gather temperature-solubility data.

Create a solubility curve using temperature-solubility data.

Predict solubilities at different temperatures using a solubility curve.

Evaluate the effect of temperature on solubility.

MATERIALS

• potassium nitrate, KNO_3, 20 g

EQUIPMENT

- beaker, 250 mL
- beaker, 400 mL
- graduated cylinder, 10 mL
- hot plate
- LabPro or CBL2 interface
- ring stand

- stirring rod
- test-tube rack
- test tube, 20 mm × 150 mm (4)
- TI graphing calculator
- utility clamp (2)
- Vernier temperature probe (2)

SAFETY

- Wear safety goggles when working around chemicals, acids, bases, flames, or heating devices. Contents under pressure may become projectiles and cause serious injury.

- If any substance gets in your eyes, notify your instructor immediately and flush your eyes with running water for at least 15 minutes.

- Secure loose clothing, and remove dangling jewelry. Do not wear open-toed shoes or sandals in the lab.

- Wear an apron or lab coat to protect your clothing when working with chemicals.

- Never return unused chemicals to the original container; follow instructions for proper disposal.

- Always use caution when working with chemicals.

- Never mix chemicals unless specifically directed to do so.

- Never taste, touch, or smell chemicals unless specifically directed to do so.

Procedure

EQUIPMENT PREPARATION

1. Obtain and wear goggles.

2. Label four test tubes 1–4. Measure out the amounts of solid shown in the second column below (amount per 5 mL) into each of these test tubes. **Note:** The third column (amount per 100 g of H_2O) is *proportional* to your measured quantity and is the amount you will enter for your graph in Step 10.

Test tube number	Amount of KNO_3 used per 5.0 mL H_2O (weigh in Step 2)	Amount of KNO_3 used per 10^2 g H_2O (use in Step 10)
1	2.0	4.0×10^2
2	4.0	8.0×10^2
3	6.0	1.2×10^3
4	8.0	1.6×10^3

Solubility and Chemical Fertilizers *continued*

3. Add precisely 5.0 mL of distilled water to each test tube (assume 1.0 g/mL for water).

4. Plug the temperature probe into Channel 1 of the LabPro or CBL2 interface. Use the link cable to connect the TI graphing calculator to the interface. Firmly press in the cable ends.

5. Turn on the calculator, and start the DATAMATE program. Press [CLEAR] to reset the program.

6. Set up the calculator and interface for the temperature probe.

   a. Select SETUP from the main screen.

   b. If the calculator displays a temperature probe in CH 1, proceed directly to Step 7. If it does not, continue with this step to set up your sensor manually.

   c. Press [ENTER] to select CH 1.

   d. Select TEMPERATURE from the SELECT SENSOR menu.

   e. Select the temperature probe you are using (in degrees Celsius) from the TEMPERATURE menu.

7. Set up the data-collection mode.

   a. To select MODE, press [▲] once and press [ENTER].

   b. Select EVENTS WITH ENTRY from the SELECT MODE menu.

   c. Select OK to return to the main screen.

8. Fill a 400 mL beaker three-fourths full of tap water. Place it on a hot plate situated on (or next to) the base of a ring stand. Heat the water bath to about 90°C, and adjust the heat to maintain the water at this temperature. Place the temperature probe in the water bath to monitor the temperature and to warm the probe. **CAUTION:** *To keep from damaging the temperature probe wire, hang it over another utility clamp pointing away from the hot plate, as shown in* **Figure 1.**

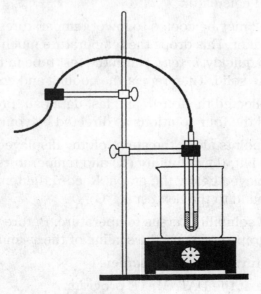

FIGURE 1

9. Use a utility clamp to fasten one of the test tubes to the ring stand. Lower the test tube into the water as shown in **Figure 1. Note:** To dissolve all of the KNO_3, you must heat test tubes 3 and 4 to a higher temperature than test tubes 1 and 2. Use your stirring rod to stir the mixture until the KNO_3 is *completely* dissolved. Do not leave the test tube in the water bath any longer than is necessary to dissolve the solid.

DATA COLLECTION

10. You are now ready to collect temperature-solubility data.

a. When the KNO_3 is completely dissolved, select START from the main screen.

b. Remove the temperature probe from the water bath, wipe it dry, and place it into the solution in the test tube.

c. Unfasten the utility clamp and test tube from the ring stand. Use the clamp to hold the test tube up to the light to look for the first sign of crystal formation. At the same time, stir the solution with a slight up and down motion of the temperature probe.

d. At the moment crystallization starts to occur, press ⌜ENTER⌝. Enter the mass (g) in the TI calculator (mass [g] is the solubility value in column 3 of Step 2, *g per 100 g H_2O*).

e. After you have saved the temperature-mass data pair, return the test tube to the test tube rack and place the temperature probe in the water bath for the next trial.

11. Repeat Steps 9 and 10 for each of the other three test tubes. Here are some suggestions to save time.

- One lab partner can be stirring the next KNO_3/water mixture until it dissolves while the other partner watches for crystallization and enters data pairs using the TI calculator.

- Test tubes 1 and 2 may be cooled to lower temperatures using cool tap water in the 250 mL beaker. This drops the temperature much faster than air. If the crystals form too quickly, *briefly* warm the test tube in the hot-water bath and redissolve the solid. Then repeat the cooling and collect the data pair.

12. When you have collected the fourth and last data pair, press ⌜STO▸⌝ to stop data collection. Discard the four solutions as directed by your instructor.

13. Examine the data points along the curve on the displayed graph. As you move the cursor right or left, the solubility (X) and temperature (Y) values of each data point are displayed below the graph. Record the temperature values in your data table (round to the nearest 0.1°C).

14. Prepare a graph of solubility versus temperature. Before you the print the graph, set up the graph style and the scaling of the *x*- and *y*-axes:

a. Press ⌜ENTER⌝ to return to the main screen.

b. Select QUIT to exit the DATAMATE program.

TI-83 and TI-83 Plus Calculators

c. To plot a graph with solubility on the vertical axis and temperature on the horizontal axis, press [2nd] [STAT PLOT] and then select Plot1. Use the arrow keys to position the cursor on each of the following Plot1 settings. Press [ENTER] to select any of the settings you change: Plot1 = On, Type = ⌐⁞⁞, Xlist = L2, Ylist = L1, and Mark = ■.

d. To scale the axes and set increments, press [WINDOW]. Scale the temperature from 0 to 100°C with increments of 10°C. Scale the solubility from 0 to 200 g with increments of 10 g.

e. Press [GRAPH] to view the graph of solubility versus temperature.

f. Print a copy of the graph of solubility versus temperature.

TI-73 Calculators

c. To plot a graph with solubility on the vertical axis and temperature on the horizontal axis, press [2nd] [PLOT] and then select Plot1. Use the arrow keys to position the cursor on each of the following Plot1 settings. Press [ENTER] to select any of the settings you change: Plot1 = On, Type = ⌐⁞⁞, Xlist = L2 (press [2nd] [STAT] and select L2), Ylist = L1 (press [2nd] [STAT] and select L1), and Mark = ■.

d. To scale the axes and set increments, press [WINDOW]. Scale the temperature from 0°C to 100°C with increments of 10°C. Scale the solubility from 0 to 200 g with increments of 10 g.

e. Press [GRAPH] to view the graph of solubility versus temperature.

f. Print a copy of the graph of solubility versus temperature.

TI-86 Calculators

c. To plot a graph with solubility on the vertical axis and temperature on the horizontal axis, press [2nd] [STAT] and select PLOT. Select PLOT1. Press [▼] to move to a new setting, and press [ENTER] or the menu keys to change a setting. Use the following settings: Plot1 = On, Type = ⌐⁞⁞ (select SCAT), Xlist Name = L2, Ylist Name = L1, and Mark = ■.

d. To rescale the x- and y-axes, press [GRAPH] and then select WIND. Rescale the temperature by entering a value of 0 for xMin, 100 for xMax, and 10 for xScl. Rescale the solubility by entering a value of 0 for yMin, 200 for yMax, and 10 for yScl.

e. Select GRAPH to view the graph of solubility versus temperature.

f. Print a copy of the graph of solubility versus temperature.

Solubility and Chemical Fertilizers *continued*

TI-89, TI-92, and TI-92 Plus Calculators

c. To view the data matrix, press [APPS], select Data/Matrix Editor, and then select Current. Press [F2] to select the Plot Setup menu. Highlight Plot 1, and press [F1] to select it. Choose Scatter for the Plot Type and then Box for the Mark. Press [c] [2] to enter the *x*-axis and [c] [1] to enter the *y*-axis. Press [ENTER] twice. **Note:** Use [ALPHA] [L] instead of [L] on a TI-89 calculator.

d. Press [♦] [WINDOW]. Scale the temperature from 0°C to 100°C with increments of 10°C. Scale the solubility from 0 to 200 g with increments of 10 g.

e. Press [♦] [GRAPH] to view the graph of solubility versus temperature.

f. Print a copy of the graph of solubility versus temperature.

DATA TABLE

Trial	Solubility (g/100 g H₂O)	Temp (°C)
1	4.0×10^2	
2	8.0×10^2	
3	1.2×10^3	
4	1.6×10^3	

ANALYSIS

1. **Analyzing Data** Using your calculator, apply a best-fit curve to the solubility graph created in Step 14. Print your graph with the best-fit curve. Label both axes, and show correct units. Label tick marks with the numerical values they represent.

2. **Analyzing Data** During fertilizer production, the concentration of potassium nitrate in the vats has been raised to 135 g per 100 g of H₂O. Based on your results, at what minimum temperature would you recommend the vats be kept to avoid crystal formation? **If the vats were not allowed to drop below 70°C, there would be no danger of crystal formation.**

3. **Analyzing Data** According to your data, how is the solubility of KNO₃ affected when the temperature of the solvent is increased? **As the temperature of the solvent increases, so does the solubility of KNO₃.**

4. Analyzing Results Using your printed graph, tell if each of these solutions would be safe to run through the factory pipes:

a. 55 g of KNO_3 in 100 g of water at 40°C <u>yes; unsaturated</u>

b. 100 g of KNO_3 in 100 g of water at 50°C <u>no; saturated</u>

c. 150 g of KNO_3 in 100 g of water at 80°C <u>yes; unsaturated</u>

d. 80 g of KNO_3 in 200 g of water at 90°C <u>yes; unsaturated</u>

5. Analyzing Information According to your graph, at what temperature would

45 g of KNO_3 completely dissolve in 100 g of water? <u>**At 30°C, ~45 g of solid**</u>

<u>**KNO_3 would dissolve in 100 g of H_2O.**</u>

Conclusions

1. Evaluating Results Research the solubility of potassium nitrate using the Internet. Compare your solubility table data to established solubility values.

How does your data compare? Explain if it is different. _____

Answers may vary according to student results and Internet sources used.

_____

_____

_____